Keeping ≈ Warm

Keeping ≈ Warm

A Sensible Guide to Heat Conservation

Peter V. Fossel

Illustrated by Jan Adkins

A GD/Perigee Book

Perigee Books
are published by
The Putnam Publishing Group
200 Madison Avenue
New York, New York 10016

Designed by Richard Oriolo

Library of Congress Cataloging in Publication Data

Fossel, Peter V.
  Keeping warm.

  Bibliography: p.
  Includes index.
  1. Dwellings—Energy conservation. 2. Dwellings— Insulation. 3.
Dwellings—Heating and ventilation.
  I. Title.
TJ163.5.D86F67   1983   697   83-8284
ISBN 0-399-50845-7

First Perigee printing, 1983

Printed in the United States of America

1   2   3   4   5   6   7   8   9

Acknowledgments

~ My thanks (if they don't already know it) to:
~ Anne Kostick, a very patient editor who knew how to point a rough idea in the right direction.

Donald Cleary of the Jane Rotrosen Agency, who got everything rolling.

Dick Ketchum and Tom Rawls of Blair & Ketchum's *Country Journal*, in which some of this material first appeared.

Mike Durham of *Americana* magazine, who published parts of the Wood Stove chapter.

Annie Proulx, who provided a wealth of information and sources on warm clothing.

Rick Schwolsky, Dave Parker, Charles Wing, William Shurcliff, Gautum Dutt and others, who gave their time and expertise and asked nothing in return.

And to St. John, the most obstinate and ill-mannered goat ever born, for generally behaving herself and destroying nothing important while this book was being written.

For Leslie,
without whom the warmest fire leaves a chill

Contents

Catching Heat
(An Introduction)

∿ When a fuel shortage was discovered crawling up
∿ the leg of North America some years ago, one of
our first official reactions was to turn down thermostats
and shut off the lights behind us. Conservation meant
sacrifice, and the idea was that somehow we could save
heat by not using any. By being cold.

Now, that's a little like trying to save money by
being poor. A lowered thermostat doesn't save heat at
all; it saves fuel. But a tank full of Number 2 oil is
nothing to curl up with on a cold winter's night.

To conserve *heat* you must trap and hold on to the
stuff where it does some good—and not let it drift outside to get the sap running in maple trees an hour before their scheduled time next March. As it happens,
by conserving heat you also save fuel, keep yourself
and your winter dwelling more snug and self-reliant,
reduce air pollution and ocean oil-spills and probably
even have some effect on U.S. foreign policy; I don't
know. But conserving heat does keep you warm, and
this book has to do with the art of keeping warm.

That doesn't mean finding exotic new sources of
fuel, but merely trapping the heat we already have.
The fires of winter burn in many places. Our prodigal

sun, for instance, throws away energy right and left, and all we have to do to stay warm is to get in its way. Cordwood produces heat and so do cookstoves, and refrigerators, pets, the land itself and even our own bodies. A furnace is the last thing you need, and many houses are doing away with them entirely and staying warmer than ever before.

Simplicity, in fact, is part of the secret.

An old friend of ours used to own a farm pickup truck which he called Number Two. It was mostly a Ford, but the only remaining evidence of that was on the steering wheel hub. The old man had discarded any pieces that rusted, rattled or broke down, reasoning that the fewer moving parts a thing has, the less time you will waste fixing it. At last look, Number Two was well on the way to being maintenance-free.

The idea made good sense, but had some limitations when applied to a pickup truck—which, after all, was *supposed* to move. Winter shelters are another matter. A house, apartment, bed or even the clothes on your back can be made to sail cozily through a North Country winter with fewer moving parts, and probably less oil, than an old farm truck, while keeping you a good deal warmer and more comfortable at the same time. As it happens, that is the whole idea of this book.

The first step is to look closely at how nature gets her creatures through a hard winter. She knows the rules.

Keeping ≈ Warm

1

The Season

~ Winter, in New England, begins leaving small
~ scratch marks on the back door by late August.
Nobody talks much about it and Barney, down in his
cluttered one-room gas station, won't start accepting
weather complaints until mid-November. The signs
are small, almost imperceptible now, but they have
begun to gather like dry leaves before a storm along
the edges of the land. One morning (the turn of a season is always clearest then) I leave the house early, letting the screen door bang shut behind, and walk down
toward the vegetable garden, past the apple trees and
an ancient sugar maple whose scars from a hundred
springtime taps have healed, and I will know before
the path takes too many turns, know without thinking
about it, that something is in the air. Some message
was left behind during the night. Or rather, there is
nothing in the air at all. No mosquitoes to slap this
morning, no horseflies, no honeybees shopping
through the corn tassels like tourists at a yard sale. The
corn, indeed, has been picked and put by for winter already, leaving behind great hybrid stalks of grass to
weather and dry. The land is pulling in, curling up its
vulnerable parts against the cold to come. Gypsy
moths, whose caterpillar larvae edited the oaks in a
voracious binge of gluttony last spring, have tucked

their eggs between crags of bark and insulated them against the winter with a stiff silken roof. I ponder scraping off these sturdy bug-houses (they say you should) but decide instead that anything so determined to survive this winter, any house that well built, should stand. The redwing blackbirds, at least, will applaud that decision next May.

But with the bugs, birds too are leaving. Generations of barn swallows, which arrive each year at lilac time to patch their mud nests under the eaves by the goat barn, feed only on flying insects—swooping and darting and complaining all summer that the bug selection is going to hell. They left last week, drawn by a secret, overpowering urge to join thousands of their kind and follow the stars of the Southern Cross five thousand miles to Brazil and Argentina; their wings and tails too long, too exposed, to last through the North Country cold. The sun's pendulum swung shorter each day, losing a minute on every arc, until it triggered some primordial trip wire in the pituitary gland and ordered these graceful companions of summer on their heroic, no-frills flight.

A similar message is written in the genetic code of my squash patch. Broad green dinner-plate leaves lay drooped and lifeless across their meandering vines this morning, and I move quickly to see what plague is robbing my winter stores. But it is only time. With their canopy gone, mature butternut squash are turning yellow-ripe under the open sun; the leaves fell merely because their work was done. They had shaded their charges from dry summer heat until a hard squash shell could form, and now the genetic keys to another generation were tucked safely away inside pounds of fibrous meat, as secure from the ravages of winter as any migrating barn swallow. Well, actually, these particular squash are headed for the root cellar, and will help insulate my own genetic code in the barren days ahead. But we will save some seeds and plant again, so two unfathomable mysteries are safe with each other.

The sugar maple, too, is calling in its summer canopy, choking off sap to the leaves and concentrating it now on the tiny buds already swelling imperceptibly at their feet. These leaves, so vital a few short weeks ago, would be deadly wicks in the dry winds of winter. Sugary sap is drawn back in toward the trunk, where early frosts convert it to starches immune to January's hammering freeze. Gray squirrels, which borrow upper branches for their summer homes—building the dry-leaf nests called dreys—now follow the route of the maple sap, abandoning these drafty shanties and searching for an empty hollow in some cozy trunk. There, wood's natural insulating ability will keep them warm and out of the stinging winds which can lift an animal's fur, or a bird's feathers, and drain its life.

But for now August is a lazy month, a time for laying in stores and building up layers of fat which will later be needed to both fuel and insulate the metabolic furnaces which can never, even for one night, be allowed to go out. Farm dogs and psychiatrists, both of whom ought to know, simply take the month off. Geese feeding in Canada's summer bogs will build up seven times their normal body weight in fat, preparing for an odyssey which dwarfs, in magnitude and splendor if not in length, the flight of the swallows. By early September, Canada geese are moving from their summer grounds in northern Quebec. From Hudson Bay and James Bay and Mackenzie Bay—from Newfoundland and Labrador—the great flocks gather. Small at first, then swelling in number, some head west while others move south following the Gulf of Saint Lawrence, joined now by those coming southeast from Ontario, filling the skies with formation after formation of great, honking leviathans whose wings beating the air can be heard a thousand feet below the ancient Atlantic flyway. Watch the stars, remember the stars. Keep the coast on your right wing. At dusk the flock appears overhead and farmers, hearing the rush of air, look up from their fields. Winter is at hand. Ice and snow have locked the northern marshes and are moving south in the wake of the geese.

The goats in our dairy barn cut back their milk production again. They sense which way the wind is blowing and are pulling in, conserving. Without our knowing it, the sun has dropped below some invisible but critical point in the September sky, its burning summer heat now turning to a gentle warmth which makes us lazy. Which *instructs* us to be lazy, to build up reserves. There is reason behind Labor Day's being when it is; and Thanksgiving, our own ritual fattening time (which, happily, follows hard upon the fattening season of turkeys, pumpkins, potatoes and the like), is a vestigial custom evolved from a medieval English feast called "Harvest Home," which lasted a week or more. We hold it in the wrong month now—it should be earlier—but then, our instincts aren't what they used to be.

The squirrels by now have stripped the oaks of their plump acorns, discarding the protective berets and burying the meat underground—turning our backyard into a shallow root cellar. They will come for them in the spring, guided unerringly to each hole by some preposterous mechanism buried deep in their brains. For now, they will sleep.

Few animals actually hibernate, actually build up

enough fat to coast through winter without raiding the cupboard. Ground squirrels do, and woodchucks and bats, but not tree squirrels. Woodchucks turn down their thermostats to 36 degrees, drop into a deathlike hibernation taking a light breath every five minutes or so and ignore whatever befalls the outside world after October. Others—the red and gray squirrels, the fox, the beaver and the bear—simply drift into a quiet stupor for days or weeks at a time, their metabolic furnaces banked and silent against the lean days ahead. They emerge only to pick through the cache of food, to hunt briefly or, in the case of the beaver, to defecate away from home. In countless lairs and hollows and niches, life gathers itself in, pulling legs to a fetal position and curling furry tails across the head to give winter the smallest target possible. The greater the surface area of an animal, or a house for that matter, the quicker it loses precious heat. Buried deep in my woodpile, the roof of a field mouse nest is low, its walls close by, hardly larger than the occupant. This mouse (this warren of mice, it turned out later) plastered its walls with dry grass and leaves and bits of fur, insulating it better than my own house once would have.

By December, snow becomes a warm blanket across the meadow and forest floor—an insulating layer of billions of tiny air pockets between the ice crystals. In nature only dry wood or peat can beat its remarkable ability to store heat. Like a soapstone hearth which stays warm long after the fire dies, the earth soaked in summer heat now gives it back to the creatures who forage across its surface. The soil, ten feet down, is warmer in January than it is in July—a fact which was not lost on those who built such deep cellars in New England houses two centuries ago. During the most bitter subzero freeze, the still air beneath a deep blanket of snow remains a balmy 32 degrees, and the red squirrel burrows snugly in here to sleep.

The weasel, in his predatory wanderings, swims below or atop this powdery snow blanket with equal ease, protected from the cold by thousands of hollow thermal hairs. A dense coat of short underfur provides insulation, while longer "guard hairs" shed water and rain. Other fur-bearing animals have the same winter protection to a greater or lesser extent, but the weasel gets to trade in a brown summer coat every December for the regal white fur which for four glorious months makes it an ermine. Same animal, different price tag.

The weasel might make it through winter in the highest style, but for pure common sense I favor the possum. One possum in particular. This matronly visitor waddled into the barn milk room in a family way one bitter night last December, clearly upset with the state of affairs outside, nestled in behind a hay bale while my wife was milking, rearranged things for a moment or two and wriggled herself in to stay. She spent the winter, living off the fat of Purina Cat Chow which she shared with the two barn cats, baring her pointed lips now and then to make it clear whose barn this was.

I envy her straightforward approach to winter, but hope that neither she nor her offspring ever discover that there is a wood stove in our kitchen.

The dew is wet on my shoes as I walk back up from the garden, my pace a bit faster than before. There is no woodsmoke in the air, no frost on the field grass yet to slow me down. The wood, in fact, hasn't even been stacked. It sits in a meandering buttress on the lumber road, unsplit and far from where it's needed. My wool shirt is worn at the elbows and is missing three buttons. There is squash to pick and chimney flues need cleaning from the previous winter. A bobwhite announces its presence from a scrub thicket across the field, but I don't have time to talk. The seasons are turning, and I have my own preparations to make.

2

Winds and Drafts

≈ Now, this is the truth. In many a drafty old New England farmhouse, they used to say the best way to tell wind direction on a winter night was to hang an iron plow chain from the bedpost and see which way it leaned. If the chain ever started swinging around horizontally, snapping off links, you could figure on a good storm blowing up before morning.

To some of that day, this was doubtless a simple matter, calling for more firewood and stronger chains. But the spread of farmlands in the early 1800s had left great swaths of New England with something of a cordwood shortage, so around 1820 the journal *New England Farmer* began suggesting ways to conserve wood and stay warm by "winterizing" the old place instead. Stuff rags into broken windowpanes, the editor advised, and paste newspaper over cracks in the wall.

Clearly, weatherproofing technology hadn't yet peaked, and "cold air infiltration" was still known as "a draft"—but the principle was there, and that simple cornerstone of a conservation notion still holds: a rag in a broken window is worth a lot of firewood (or fuel oil).

In effect, the better a house can *store* its heat, the less you have to buy. Even the most sophisticated new energy conservation measures, from insulated window shades to automatic flue dampers, operate on this straightforward idea. Rag fibers are simply rolled into felt weather stripping today, while newspaper is shredded and flameproofed for cellulose insulation. When it comes to cold farmhouses, scientists and farmers think alike.

Within the last several years, with the use of common weatherproofing techniques, common sense and a dash of passive solar heat (south-facing windows), new houses have been built to trap and store warmth so effectively that 70-degree indoor temperatures can be maintained through a howling North Country winter for the tantalizing cost of about $100 in fuel oil or gas. The next question is: Can the estimated 80 million existing houses built in this country since the late 1600s, and the untold number of city apartments, be retrofitted—shored up and plugged—to achieve the same high degree of energy efficiency?

Nope. Probably not. But, depending on site and design, most can come darn close. Proper insulation and tightening alone can reduce the annual cost of heating a typical North country house by 50 or 60 percent according to federal energy officials, and that can easily mean $700 to the average homeowner in many states. Other measures, some of which cost little or nothing, can make a place even more snug and heat-efficient, and it's not difficult to retrofit even an old barge of a house into something that can sail through a hard winter on three or four cords of firewood.

Construction and mortgage costs aside, there are several nice things about retrofitting a house rather than building anew. For one, most of the work can be done yourself, and it needn't be done all at once. For another, the fact that the house is already built narrows the options which have to be juggled and studied. (The wall cavities will hold about 4 inches of insulation, and that's that.) Finally, it is rewarding work that increases the comfort, value and life of an old place.

The first step in converting any dwelling into a warm and self-reliant winter home should be a thorough energy audit. Which, I suppose, is a nasty choice of words. Having your house "audited" is like inviting the garden club to inspect your socks after a five-day fishing trip. Nonetheless, it is a near foolproof way to calculate how you can prevent the most heat loss for the smallest or shortest-term investment. Virtually all weatherproofing measures are free in that lower fuel bills pay for them and then they just sit there capturing warmth long afterward. But some earn their keep quicker than others. In snow country, for example, weather stripping and attic insulation generally pay for themselves the first winter whereas blown-in wall insulation is apt to take several heating seasons. A good energy audit helps determine the best retrofit options for a particular house, and lists the payback period for each—which can vary tremendously depending on local climate, the cost of your heating fuel and a batch of other factors. It's a battle plan—like sketching out the vegetable garden at the kitchen table before churning off into the lower forty with an armful of seeds.

The federal government's Residential Conservation Service requires large utility companies to offer customers home audits for $15, and some of these are quite sophisticated, feeding information into a computer which reads back recommendations over a special transmitter hooked up to your telephone. Other audits are . . . well, less technical. A well-meaning employee might simply walk through the house, asking questions and confirming what you already suspected.

And some audits are flat wrong. Princeton University researchers a few years ago found that heat loss through the insulated attics of forty typical homes was three to five times higher than a conventional audit would have predicted—not because the insulation wasn't there, but because escaping heat paid no attention to it and drifted up through cracks around the chimney, stovepipe, plumbing vent, electrical wiring and attic doors. It escaped through gaps where insulation wasn't tucked snugly up against attic floor joists.

Winter can chill a house the way seawater creeps into a wooden boat—through the small places, where the toughest oaken planks don't meet.

The biggest drawback to a professional energy audit, however, is that it is based on standards and averages which apply to a "typical" dwelling, but not your own. They don't allow for the bumps and quirks and odd angles and family habits which give every home its character and make it a special case. Nor can they compute for ingenuity and inventiveness, which so often mark not only nature's trek into winter but our own as well. Rick Schwolsky, a quiet but determined solar energy expert from Grafton, Vermont, who installed solar panels on the White House, once converted a heat-wasting exterior-wall chimney, which everyone knew couldn't be insulated, into a heat-storing thermal mass by enclosing its brick base in a south-facing solar greenhouse.

A more subtle flaw in professional energy audits is that once we lean too heavily on experts to tell us what is wrong with our homes, we find ourselves needing other experts to fix it.

Instead, become one. A house is not a complicated thing. Learn the secrets of the place. Poke and probe and measure; see how it was built and what was used.

Crawl into crawl spaces and pry up a board on the attic floor to find what is under it. Learn where the house gains and stores heat, and where it loses it. Unscrew a few electrical switch plates on the outer walls to look for insulation. Track where the plumbing and heating pipes run, and note in your mind every small break in the walls or roof—such as where utility wires enter or hot air leaves from the dryer vent. Watch the sun on its low arc across the winter sky and know where the prevailing winds blow from—and what is there to block them both. Even in an old farmhouse the sun can replace an oil furnace as the primary source of heat; but an unobstructed wind can rob a house of its entire warm air supply in a matter of minutes.

Open damper in chimney
Uninsulated walls
Poorly insulated attic
Leaky windows
Cracks along top of foundation
Uninsulated basement

Illustrations by Jan Adkins
Air Leaks Causing Heat Loss

An understanding of how your house moves through winter, and the most cost-effective ways to improve it, takes observation and detective work. The knowledge builds with time. One crack leads to another, and the solution to a particularly frustrating problem—such as how to insulate a cathedral ceiling without hiding exposed rafters—can lie fallow for months in the back of your mind before a solution sprouts.

In retrofitting any house, the first place to start is with drafts. Most heat losses occur through simple air leaks, which can account for anywhere from 20 to 50 percent of an annual heating-fuel bill. In fact, most houses lose their entire volume of warm, indoor air every thirty to sixty minutes because of gaps and holes in the building envelope. Some are tiny, some nearly inaccessible, but they all add up. A 1/16-inch crack around a doorframe is equal to a hole the size of your fist, and the effect of all such leaks in even a well-insulated house is about the same as if you left a double-hung window open all winter. Actually, it's far worse, because two small holes set up a draft, whereas one large hole doesn't. Animals adapted to that principle long ago in building their burrows.

Of course, some fresh air is necessary for health, and to provide a draft for the furnace, wood stove or whatever, but it's a rare house which can be retrofitted to the level where tightness is a problem. In any case, a window can always be opened and the air infiltration controlled.

The keys to plugging air leaks in a house are patience and attention to detail, and the basic materials are caulking compounds and weather stripping. These are relatively cheap and generally earn their keep before the spring thaw. It is important, however, to tighten only the *inside* surfaces of a house. Otherwise, moisture condensation may become trapped in the wall cavities, reducing the effectiveness of insulation and laying a welcome mat for dry rot. (More on all that later.) The outside of a house should be tightened and caulked only to the extent necessary to keep out rain, bugs, squirrels and the like. Exterior sheathing, shingles or clapboards don't have a high enough insulating value to warrant more tightening than that.

The best time to caulk indoors is when your house is dry, the wood has shrunk and cracks have opened as wide as they are going to. In northern climates this usually means from early fall to winter, because cold air holds little moisture and dries out a house. Good caulk will cost $2 or more per tube (enough for a ¼-inch bead about 25 feet long), but it adheres better to the sides of an opening, shrinks less, remains flexible longer to move with surfaces which swell and shrink with humidity changes and can last for decades. By comparison, some oil-based caulks harden and crack in a year or so. You can find yourself caulking the caulk.

The best directions to follow are those on the tube. The only thing they never mention is to carry a few paper towels in your back pocket—I have yet to see a caulking compound that stops squirting when the plunger is released on the caulking gun. Maybe they do for other people. It is hard to keep unused caulk from hardening in the tip of the tube when being stored, but sometimes a nail stuck in the hole can help. Caulk isn't that expensive that it is worth saving for-ever.

Acrylic/latex caulk is the best general-purpose type and is excellent for plugging gaps in wood, plaster and drywall. It sticks to damp surfaces, can be cleaned up with water, holds paint well and should last for up to ten years. The more durable, flexible and costly sili-cone, polysulfide and neoprene compounds should be used in a seam where there is movement between two surfaces because of stress (a slamming door) or humid-ity changes (between pine paneling boards or in floor-board joints). These require special thinners for cleaning up, however. They can last for up to twenty years. Silicone makes an excellent clear, colorless caulk which cannot be painted but is nearly invisible in a small seam.

Caulking works best in gaps less than a ¼-inch wide and should be forced deep into the opening, not just spread on the surface. Even hairline cracks should be filled. Acrylic/latex caulk holds up surprisingly well in the cracks between wide floorboards, despite movement from shrinkage and swelling, and comes in a variety of brown colors for use with unpainted wood. Wide gaps between floorboards should be filled with a strip of the same type of wood, which can be held in place with a bead of brown caulk laid down beforehand.

Large holes or cracks hidden from daily view, such as those you might find around the sill in an old cellar, can be stuffed with oakum—an oily, fibrous, ropelike substance used for caulking wooden boat hulls.

I've never actually seen oakum, and don't know anybody who has, but it's recommended in the newspapers all the time. I asked for some once while buying roofing nails in a small New Hampshire town, but the clerk glared at me and declared, "this is a *hardware* store . . ." As if somebody my age shouldn't be using the stuff in the first place. It's probably wise to wait until the local lumberyard puts up a sign saying "Oakum sale," and then drop in.

Better yet, use fiberglass insulation. Pieces from a fiberglass batt or roll can be tucked snugly into almost any size hole. Tight packing eliminates the insulating value of this material by squeezing out its tiny air spaces, but for plugging air leaks this isn't that important. Urethane foam, which comes in aerosol cans, is often a more permanent solution for filling these large gaps, and works especially well in a rough stone foundation or around sill plates or basement windows and doors where gaps are hard to reach. Use the cans that come with an extension nozzle. Urethane is expensive and sometimes messy to apply, but it serves as a vapor barrier, insulation and powerful adhesive. It is also flammable and gives off highly toxic fumes when burning, so don't use it near heat sources or above the foundation level. Figure out exactly where it's going to go and try to use the entire can at once because the material can harden and clog the nozzle.

Urethane deteriorates in sunlight and isn't handsome to look at, so fill gaps in the outside of a foundation by repointing them with mortar. This isn't difficult, and is a nice art to learn if there is other masonry in the house. Or if you wish there were.

Probably the hardest part of sealing air leaks inside a house is the detective work required to find them. The obvious spots are where walls meet ceilings, floors, or window and door casings. Look under sinks, in the back of closets and cabinets, behind appliances on an exterior wall and behind pictures for old nail holes.

If you never lowered the top half of a double-hung window last summer (deliberately, at least), consider caulking it shut. Caulk them all shut. The summer air circulation which results from opening the top and bottom of a double-hung window is negligible, except in a closed room with only one window.

In brief, caulk the seams where any two fixed materials meet in the house. Some of the worst cold drafts occur where a wall or mopboard joins the floor. Seal the joints above and below metal baseboard heating units, as well as the seams where they join. Holes where the hot water heating pipes are cut into walls or floors require a caulk such as silicone that adheres to metal, stays flexible and can stand high temperatures. Caulking these spots is especially important if the walls have not yet been insulated, because cold air in the wall cavities will sink to the bottom and pour across the floor at an enormous rate.

Some of the greatest heat losses in a house are from around hot air heating ducts, foundation soleplates and electrical outlets in an exterior wall. In fact, these outlets and wall switches account for 20 percent of the air leaking through a typical new house—equal to that lost through windows and doors combined according to one field study. Inexpensive foam gaskets are made which fit snugly behind the faceplates of these outlets and switches. Even more air (23 percent) leaks in between the foundation and wall, according to that study.

Cold air can also pour inside through hollow interior partitions which act as a tunnel to an outside wall or attic. While tightening around these partitions can lead to more work than most of us are ready for, it's worth keeping in mind next time you wonder why the closet is so cold.

Perhaps the most important consideration of all in tightening an old house is the so-called stack effect, which pulls air from a basement up through the house and out the attic—unless you stop it. This makes a tight first floor and attic floor vital. In the attic, plug the openings where pipes and wires enter from below with chunks of unfaced fiberglass insulation, even if it means digging under attic floor insulation to get at the holes. Fire codes often require that insulation be kept 3 inches from recessed lighting fixtures to prevent overheating and fires, but the gap around a chimney can be stuffed with unfaced fiberglass. Researchers for Princeton University's Center for Energy and Environmental Studies recently found that warm air loss through leaks into the attics of wood-frame houses account for one-fifth of all residential space heating energy use in the country—and about 2 percent of *all* energy use. So tighten the floor of an unheated attic better than any other surface of the house.

Caulk and plug also around dryer, stove and bathroom vents; places where pipes and wires enter outside walls from the basement; and all the inside seams on window and door casings. There will be dozens of other spots, varying with the house, and tracking down noticeable drafts will only lead you to the larger ones. This can be a frustrating, rewarding, damnable and fascinating job which will teach you a good deal about houses and building methods, and make the place look and feel more snug than you might have thought possible.

The next step is to weather-strip windows and doors. If they already are weather-stripped, check for drafts anyway. Old weather stripping can wear out or move with time and banging doors, and the most exotic weather stripping is worthless if it doesn't provide a tight seal. For retrofit work, spring-metal strips are

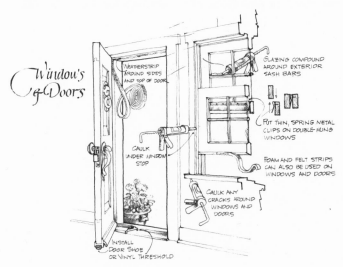

Windows & Doors

WEATHERSTRIP AROUND SIDES AND TOP OF DOOR

GLAZING COMPOUND AROUND EXTERIOR SASH BARS

PUT THIN, SPRING METAL CLIPS ON DOUBLE-HUNG WINDOWS

CAULK UNDER WINDOW STOP

FOAM AND FELT STRIPS CAN ALSO BE USED ON WINDOWS AND DOORS

CAULK ANY CRACKS AROUND WINDOWS AND DOORS

INSTALL DOOR SHOE OR VINYL THRESHOLD

usually the tightest and most durable, but in old houses with warped or dented wood and bumpy edges, soft rubber tubing, spongy foam or felt strips may be better.

With some weather stripping you can get a snug fit around doors by laying the cut pieces over a bead of adhesive caulk with the door slightly ajar. When the caulk has dried a little, close the door gently and tack or staple the weather stripping in place. Most types of weather stripping can be painted, and the materials should pay for themselves in a year or two. Felt strips are the cheapest, but wear out quickest. When buying metal strips, try to find it in 10-foot lengths; it will be cheaper per foot than in a kit and you will waste less if doing several doors and windows.

Rope caulk, such as Mortite, is an excellent temporary measure, but it tends to harden and loosen in winter, falling back out of the cracks. It also has a tendency to be rolled every spring into stringy, sticky balls which end up on closet shelves and workbenches, to be stored there forever.

If some rooms aren't heated during the winter, consider weather-stripping the doors to those areas. Thin strips of foam are the most unobtrusive thing to use on these doors, and they give them a nice snug sound when they are shut. Also tighten the doors leading to the attic or basement.

A significant amount of cold air can sneak in through the joints between window sash bars and glass if the putty is cracked and loose. Fortunately, glazing compound now comes in a caulking cartridge with a special square, flat tip to fit an outside sash bar and this saves hours of puttying old wooden sashes. If any new windows are going to be installed during the retrofit process, it's worth knowing that new double-hung wooden windows or metal casement windows leak five times as much heat as a double-hung wood/plastic window (where the wood sash is sheathed in a film of polyvinyl chloride), and ten times as much as a *fixed* wood/plastic window, according to the National Bureau of Standards. Aluminum jalousie windows in a cold climate amount to little more than horizontal holes in the oil tank. A good aesthetic and cost compromise is to seal the upper sash of existing double-hung windows, using a flexible caulk, as mentioned earlier.

There are no secrets to any of this. Use your good sense and intuition. Spend some time at the hardware store during the oakum sale, and see what new materials are available. They change every now and then. Pick those that should last as long as the house, if possible. As though you would be handing it down to your grandchildren someday. Little bits of pride and knuckle can be left in many places, and a tight door is not the least of them.

3

Moisture

≈ Cellar holes are empty places. I come across one now and then and climb down in, as though by moving my hand across the worn rock walls, green with moss and lichens now, I will be told the story of who lived here and why they left and what happened. Part of the wall has collapsed under the silent push of a hundred winter freezes and thaws, but the front step is still there . . . waiting.

My brother and I used to find these great rock-filled yawns in the land along the back roads of Hillsboro County, New Hampshire. He was looking for old bricks, which were too expensive to buy, and he could sniff out a cellar hole the way a man finds his way to the bathroom at night in the dark. A bend in the road, a break in the wall or two big "bride and groom" trees standing off in an alder thicket where they didn't belong told him that somebody lived here once.

And they had. We would stand, staring down mute at an old brick chimney collapsed in rubble and buried in lilacs or saplings. (The women planted lilacs, which sometimes spread to fill the hole after they were gone.) But what sort of leviathans—what men or oxen—had moved these giant stones from the field and laid them one upon two others so square and strong in the ground? Some young farmer, perhaps, walking up from Massachusetts, who found this fertile piece of land one April afternoon, with the stream nearby. Close enough for hauling water. A stand of white pine here for boards, and chestnut trees for beams and a mill already cutting in the village. Some farmer who, needing a house and not knowing any better, built one.

Then it burned down one winter. Or it rotted slowly, and they moved on. One or the other.

Charring acts as a natural preservative for wood, and half-burnt timbers sometimes last for generations where they fell. Moisture works more slowly but leaves fewer tracks in a cellar hole. In those earlier times, most moisture damage was caused by rain leaking in, or ground dampness which rotted sills until a house began to settle in upon itself. Today the problem is more insidious, because it is also apt to occur in the form of moisture condensation in wall cavities, where its effects are not quickly seen. This is a modern problem, caused partly by insulation, but also by the tremendous amount of water we use in our homes and the high temperatures at which we keep them.

Water condensation in a wall not only peels paint and reduces the effectiveness of cellulose and fiberglass insulation, but is a welcome mat to dry rot and termites—which both thrive in a warm, damp, dark place. Moreover, the better a house is insulated and tightened, the more serious the moisture problems can be. Unless you take precautions.

In virtually all major retrofitting work, from insulating crawl spaces to caulking, an awareness of water vapor and how it works is as valuable as boots at night in a cow pasture. It not only helps preserve your house for another generation, but helps cure dry throats and runny noses in the process.

It works this way:

Oddly enough, the dry, Sahara-like conditions of an old North Country house in winter—which cracks our lips, dries out sinuses and makes the furniture fall apart—is not caused by the furnace or wood stove drying out the air, as popular notion would have it. The dryness comes from outdoors, and the colder the weather, the worse the problem gets. Most houses lose their entire volume of warm, indoor air every thirty to sixty minutes through cracks, holes, chimney flues and so on, and in a rattly old farmhouse the air exchange rate is even worse. At the same time, cold air can hold less moisture than warm air. So when this cold, dry air leaks indoors and is warmed up, its relative humidity plummets—which means it can now hold far more moisture than it brought in. It gets hungry for water, and draws it from your skin, throat, eyes, chairs or whatever.

And that's it.

The answer to a dry house is not to buy a humidifier, but to tighten the place against cold-air infiltration.

Lesson two: the average family of four puts 3 to 5 *gallons* of water into the house's air each day through cooking, breathing, bathing, dishwashing and the like. This water vapor penetrates into unprotected attics and wall cavities, then condenses when it hits a cold surface and reaches dew point (the cold air cannot carry that much moisture). In drafty old farmhouses, the condensation simply froze unseen on rafters and sheathing, evaporating harmlessly through clapboard and shingle cracks when warm weather returned in spring. Actually, most old houses were cool enough, and used little enough water, that moisture hardly condensed in the walls and attic at all. In a modern, insulated house, however, vapor condensation will occur within the insulation itself; and the small dead-air spaces which make insulation so effective also prevent moisture from evaporating easily in warm weather, *especially* if the house has been tightened up outside. In extreme cases, cellulose insulation could turn to mush or the trapped water could sit there for months nurturing dry rot (which requires temperatures above 50 degrees F).

The solution is a vapor barrier. This is installed on the *warm* side of insulation to block relatively moist room air from escaping into the wall cavity. In effect, the only air reaching wall insulation should come from outdoors.

Two of the most commonly used vapor barriers are (1) the foil backing on batts and rolls of fiberglass insulation and (2) clear, 4-mil polyethylene plastic sheets. These are generally used only in new construction, however, or when the walls of an old house are gutted for one reason or another to expose the frame. Then the foil or polyethylene can be stapled to the studs and fitted snugly around electrical outlet boxes in a continuous sheet—which is covered with new Sheetrock or lath and plaster. In retrofit work, where insulation is blown into a wall cavity, common vapor barriers include (1) a layer of special vapor-barrier paint, (2) several layers of common wall paint, especially oil- or alkyd-based varieties and (3) vinyl wallcoverings (they call them wallcoverings these days). All of these, of course, go on the warm side of the wall.

As it turns out, however, by far the most effective way to stop vapor condensation problems (and cure scratchy throats in the process) is to seal and tighten every ceiling, floor and wall surface *on the warm side of the insulation*. It is commonly thought that the most critical function of a vapor barrier, such as polyethylene, is to act as a virtually impervious barrier to water vapor—which, by diffusion pressure, can eventually work its way through plaster or Sheetrock. But these vapor barriers also happen to be impervious to air— and this, it turns out now, may be their most valuable quality.

Two recent studies, one by the Center for Energy and Environmental Studies at Princeton University, indicate that moisture travel *through* unbroken wall and ceiling surfaces is negligible compared to water vapor *carried into* a wall cavity by air movement through cracks, open nail holes, electrical fixtures and other similar routes. This does not mean that vapor barriers aren't necessary; it means that the best way to create one is simply to block all openings through which warm room air can reach the insulation. After that, additional barriers such as vinyl wallpaper or several coats of old paint provide insurance.

The strongest vapor pressure on an outside wall is apt to occur in especially damp rooms such as a kitchen or bathroom, and it is wise to check every summer for evidence of moisture in the wall cavities. The clearest evidence is exterior paint which peels off a year or two after it is applied. Unpainted shingles or wood clapboards which curl outward may also be a sign that moisture is swelling their back sides. First check for rain leaks into the wall, then see if things can't be tightened up more inside.

One side effect of vapor barriers (beyond saving fuel and eliminating many drafts) is that they increase the indoor humidity, which is nice for furniture and sinuses. Once a house is tight, the moisture being poured into its air through cooking and dishwashing is apt to linger awhile. Then something is needed to keep this moisture away from the wall insulation. In a way, vapor barriers correct a problem they help to create.

Of course, one of the simplest and most energy-efficient ways to prevent vapor condensation problems is to keep the house cooler. This not only saves fuel but helps equalize the vapor pressure on exterior walls. It also makes your nose and throat more comfortable (again) because, even though the same absolute amount of moisture will be in the air, the *relative* humidity will be higher (meaning the cool air holds more water relative to what it's able to hold at that temperature before saturation), so evaporation from room and body surfaces will be much less. This is the reason your throat feels better outdoors than in.

Unfortunately, while it is a relatively straightforward matter to keep moisture out of wall cavities, field tests have shown that no amount of foil-backed insulation or leak plugging can keep water vapor out of an attic. This is because of the strong stack effect in a house, which pulls warm air upward to concentrate its force on the highest ceiling. The only solution here, after sealing every hole you can find, is outside ventilation.

And attic vents are ugly. Especially in a hip-roof house where there are no gable ends to slide vertical louver vents into. In no time flat, by following building rules, you can get a historic old place looking like the top of a fast-food restaurant. Of course, cellar holes aren't much to look at either, so the best advice is to mix technical rules with common sense and ingenuity. In fact, many old houses already have attic vents under the roof eaves. They are called "cracks." They work just as well as anything you buy in a store.

The usual rule for attic ventilation is 1 square foot of

open vent space for every 300 square feet of attic floor area, if there is a vapor barrier below; double the vent space if there isn't. Premade vents can be purchased and installed on roof peaks, in gable ends, on open roof pitches or under the eaves (called soffit vents). Vents on the roof peak are as unobtrusive as any, because they can be made to look like a couple of 1-by-6-inch boards such as you might find on the peak of any wood shingle roof. Keep in mind that attic air should be able to circulate either from one end of the attic to the other, or from the bottom up—and that while screening stops wasps and squirrels, it also slightly restricts air flow through a vent.

The best thing to do is to figure out how much vent space you should have according to the formula, then adjust it according to the tightness of your ceiling, the amount of vapor in your indoor air and similar factors. It doesn't hurt to actually wander upstairs now and then on a cold winter night and check for condensation on the rafters. You might find frost, dampness or dark streaks where the moisture has soaked in.

Outside walls should also allow trapped moisture to escape while shedding wind, rain and insects. A wall made of board sheathing, moisture-permeable builder's paper and cedar shingles or clapboards is excellent for this. It not only "breathes," but provides some insulation value. Layers of oil-based paint and well-caulked joints on a clapboard wall can turn breathing wood into a vapor barrier on the wrong side of the insulation; and paint that was applied properly but still peels and blisters is a sign of moisture within a wall. Again, the idea here is to allow only outside air to enter a wall cavity (preferably slowly), but not leak around the insulation; and the common practice of caulking and sealing exterior wall surfaces beyond what is required for rain protection is no longer thought to hold water. Or rather, it holds too much.

Vinyl and aluminum siding also have a reputation as wrong-side vapor barriers, but tiny ventilation holes called "weep holes" on the underside of clapboards, or cracks at the end of clapboards, should provide ventilation. Builder's foil, when used over the exterior sheathing of a house, should be perforated or slashed every foot or so; and you can make exterior-grade plywood breathe by installing ventilating plugs, if necessary, at the top and bottom of every wall space. Holes drilled for blown-in insulation should, of course, be left open behind the shingle or clapboard.

If condensation occurs behind a tightly sealed wood clapboard exterior wall (as evidenced by peeling paint, for instance), the horizontal seams between every few clapboards can be gently pried apart and wedged open slightly with narrow strips cut from a cedar shingle. Better yet, keep moisture from reaching the wall cavity in the first place.

The hitch in all of this, of course, is that not many of us can afford the time or money to insulate, caulk, seal, tighten and vaporproof our walls and attic this year. In that case—and that is the case—start slowly and tighten the walls against moisture and air. Back up if you must. Let the wall insulation wait awhile. Do the attic and the big gaps.

Meanwhile, if the house is dry, distribute as much humidity as possible throughout the house, and don't allow it to be trapped in the kitchen or bathroom. Drying firewood, open boiling pots and house plants with big, shiny leaves add a great amount of moisture to the air. Remember that now, and eliminate those your heating system. For advice, a good bet is to find out whether your local utility (the one that supplies your heating fuel) offers low-cost energy audits, and have one done. Whether or not the audit itself is worthwhile, it is an inexpensive source of information on how to adjust or shut down all or part of your central heating system for the winter, without paying for a costly service call. Utility company auditors are generally experts on central heat furnaces, are seldom in a hurry and are happy to answer questions—even if they consider your plans a bit odd.

With baseboard hot water heat, learn how to shut down and drain a zone and reopen it again. This enables you to shut off heat to one zone of the house—the upstairs, say—for part or all of the winter, and is a good thing to know in case a heating pipe freezes and the repairer can't come until the following Tuesday. Forced hot air and electric baseboard systems are immune to subzero cold, but when a hot water system is turned down to 45 degrees at the thermostat, there is a chance of freezing the heat pipes during a severe cold snap. In a drafty old house, frigid outdoor air can sneak in around the sill plate or down through an uninsulated wall cavity, and if room air isn't warm enough to counteract these drafts, they can freeze and split a hot water heating pipe. With baseboard heat, this is most apt to occur where the copper tubing has no metallic "fins," such as at the end of a baseboard unit—especially where the tubing runs through a partition or down through a floor. The fins not only dissipate heat into the room from the hot pipe, but draw relatively warm room air back into the pipe when the furnace is off. Check these sections of exposed copper tubing meticulously for incoming cold drafts, and seal them. For insurance, you might turn the thermostat up a bit during extremely cold weather.

At the same time, ask the energy auditor about installing a smaller nozzle on your oil burner. This keeps the furnace running more steadily—and therefore more efficiently—for the same amount of fuel, and also helps keep a steady supply of warm water in the pipes to prevent freezing. Most furnaces are oversized anyway. If you have weatherproofed or insulated the house at all since the furnace was installed, or keep room temperatures below 75 degrees, the furnace is almost certainly too big. This means it needs to run only sporadically, which has the same effect on your oil bill that stop-and-go city driving has on a car's gas mileage. Ideally, the oil burner should run almost constantly on the coldest night of the year. If it doesn't, then ask your oil dealer if the furnace can be fitted with a smaller nozzle. It reduces the amount of oil sprayed into the combustion chamber (which usually ranges from 1 to 3 gallons per hour) and can trim your oil bill by 10 to 20 percent. The changeover is relatively simple and inexpensive, especially if it's done while the furnace is being cleaned or adjusted, as it should be every year.

Another trick is to find out how to adjust manually the upper and lower temperature limit controls on the burner. These govern water temperature in the pipes (usually 160 to 180 degrees in a northern climate) and may be adjusted to a lower setting. Check with a furnace expert, though, to see what effect that will have on the efficiency of your particular system.

Tap-water pipes are also subject to freezing in an unusually cold house, so insulate these heavily wherever they may be exposed to the cold, and keep drafts off them. Or simply drain any pipes which aren't needed in the winter. When water in a pipe freezes it expands and can split copper tubing with ease, causing potentially nasty damage. A full basement generally stays well above freezing if it is not drafty, even in northern parts of the country, because of the earth's own stored heat and insulating capacity—but crawl spaces are less secure from cold snaps.

In any case, pipe insulation saves heat in hot water pipes, and keeps cold water pipes from sweating in summer. Cold water, whether it comes from a well or a municipal water company, is not apt to enter the house at much less than 50 degrees, so insulation will also help keep the cold water pipes from freezing in winter. A good understanding of the house will reveal where insulation is required. Hot tap-water pipes need insulation for the first 5 or 6 feet, where they draw heat up out of the hot water tank like a wick. Hot water pipes for space heat should be insulated for their entire length through an unheated basement, however, as should forced hot-air ducts. Up to a third of all heat produced in a forced air system is lost through exposed ducts in the basement, according to one recent study. Pipe insulation is expensive, so shop around and look for spring clearance sales. In some cases, a "heat tape" cord, which wraps around a pipe and keeps it warm for very little electricity, may be the best bet where frozen pipes are the major concern. Simply plug it in on cold nights.

One nice thing about all of this is that it helps keep the family safe and warm through the hardest winter, no matter what the thermostat reads. And if the furnace gives out one December night, and they say it's worn out past fixing, then never mind. The damn thing was just burning oil anyway.

9

Apartments

∿ The old skunk which sometimes clambers into the shelter of our barn to escape the hardest winter nights seems to hold me personally responsible for the small miseries of her life. She blames me for the deep snows, for the February rains that come driving in from the northwest to soak her bed under the tool shed—and even, I think, for the dwindling grub supply in that rotten old stump behind the day lilies where she roots by night. She prefers all this to my company, though (a feeling we share), and during evening milking she will emerge from some shadowy corner and waddle across the barn floor with pained deliberateness, leveling accusing glances at me all the way, headed outside to face whatever wretched burdens I have assembled for her this time. She wriggles out between two boards in the goat pen like a plump matron struggling into her girdle, and is gone.

Watching her, I wonder if I am nothing but a tenant here myself, who scarcely controls the land at all. Without my permission, the pea crop can wither in a May hot spell, and thistles will bloom where I never planted any. By law we own the barn perhaps, but nature can reclaim it whenever it chooses. In fact, judging by the lean of the timbers and shape of the sills, nature has had an eye on this particular structure for some time. And when night hurricane winds from a winter blizzard come screaming in off the coastal waters, peeling screens from our back porch like pages off a calendar, it reminds me then whose house this is. We bought the place, it seems, from another man who didn't own it either.

Which brings up the matter of landlords and rent.

Winter is an egalitarian thing, that pays no attention to names on a deed. The bleak winds that gnaw at the clapboards of an up-country farmhouse can chill a city apartment with equal ease, and most of the same rules for coping with the cold still apply. Those who rent their dwellings simply have an extra middleman between themselves and the real landlord.

Still, when it comes to snugging up a place, there are problems peculiar to apartments. For one thing, you may not plan to live there long enough to justify the price of weatherproofing. Moreover, any heat-conservation steps you take may have no effect on the rent at all. Landlords often see little sense in saving energy when their higher fuel costs can simply be passed on to tenants. Finally, on a raw November day with the prospect of winter still ahead, a city can be the coldest place on earth. Metropolitan areas register as "hot spots" on infrared satellite photos, but those who live there don't believe that nonsense for a minute.

The first step in keeping winter out of a city apartment is to remember that the place is home—and that heat conservation means comfort, not sacrifice. If cold air is leaking in through a window and rolling across the floor, then weather-strip the thing and consider buying a plexiglass storm-window kit to mount on the inside. Hallways are often unheated, so weather-strip the door at the same time. Never mind payback calculations—you will be warmer, and that is the important thing. Especially if there is not as much heat in the building as you would like.

In some cities, such as New York, good apartments remain available for about as long as it takes somebody to open a checkbook, but if you have any choice in the matter look for a place on a middle floor. Sandwiching yourself between other apartments provides winter insulation, and the top and bottom floors of a building are usually the coldest. Carpeting and pads can block the cold of a drafty basement, but won't help in a top-floor apartment. Without insulation, these tend to be cold in winter and hot in summer—especially hot under a flat asphalt roof.

Sunny windows provide free warmth, but protect them from night heat loss with heavy, insulating drapes that can be taken down when you move. These will also help eliminate floor drafts on a bitter night. If possible, choose an east-window exposure over a west-facing one, because the apartment will stay cooler in summer.

Wall hangings, radiator reflectors which direct heat into a room and couches and beds located against inside walls also help increase comfort in a winter apartment. As in any home, look for drafts and air leaks around windows, doors and wherever heating pipes and wiring enter a room, and caulk them. Kitchen and bathroom vents are notorious paths of heat loss, and can do more than anything—except perhaps an open window or fireplace damper—to drain warmth from a room. If these vents pull up air continuously, try rigging a small door with spring hinges and a pull cord to open or close the vent at will. (It is usually located near the ceiling, and out of reach.) A small piece of wood or sturdy fiberboard with adhesive foam weather stripping works fine.

This and other permanent weatherproofing measures (lasting at least three years) are eligible for federal energy tax-credits regardless of whether a tenant
or homeowner pays for them.

For serious comfort and heat savings, however, the
entire building should be tucked in for winter as
though it were just another North Country house. A
group of tenants cooperating with the owners can ac-
complish energy-conservation feats that will dwarf
anything one occupant can do, and the results not only
increase the value of a building but can reduce rents. In
Boston, for example, about 35 to 40 percent of a typical
apartment rent pays for energy consumption—for
heat, hot water and electricity—and the tenants of one
city apartment building cut about $50 off their annual
rents by holding a Saturday afternoon "weatherproofing bee" in the building's drafty old basement. Future
rent increases will lag well behind rising oil prices, and
tenants regarded the place as more of a home after the
project.

Old city apartment buildings are ideal candidates for
retrofitting because most of them have never been
insulated or tightened at all, and the work is straightforward. More importantly, large-scale energy-conservation measures—rooftop solar hot-water heaters,
heat-recovery units that produce free hot water from
the residual heat of central air conditioners, furnace
flue dampers to keep warm air from escaping when the
furnace is off—are more economical and efficient here
than they might be in a single-family house.

One large Cambridge, Massachusetts, research
firm has cut the winter heating bills for its six-story,
100,000-square-foot office building to a piddling $35 a
month by setting up a "thermal recovery system" that
recycles warmth from such sources as electric lights,

computers, refrigerators, human beings and air conditioners.

Apartment buildings have far less exposed surface area per square foot of living space than a house (only one roof and basement, remember), and also contain countless sources of residual heat which can be recycled, or at least captured, from occupants, appliances, electric lights and the like. Attic and basement insulation become more cost-effective when the living area between these two warm, sandwich layers is increased—as in a multistory building—and the same principle holds for front storm doors, pipe and duct insulation and other weatherproofing measures which don't have to be applied to each individual apartment.

Many old brick or brownstone buildings have no wall insulation, and no open wall cavities in which to put it, but this can be solved the same way it would be in a brick house—by covering exterior walls with a new, insulated stud wall on the inside, and building out window and door casings to meet the new surface. Materials for such improvements are often cheaper when purchased in bulk than by individual tenants.

A thorough energy audit can reveal scores of other ways to capture heat in a city apartment building. While tenants benefit from added winter comfort and—in all likelihood—lowered rents following in the steps of lower fuel bills, landlords, for their part, won't be blind to the advantages of permanently lowering their building's operating costs, and increasing its value at the same time. In some cases, heating cost alone is the pivot point upon which owners decide whether to restore or abandon an old building.

Now, back to earth for a minute.

The initial cost of weatherproofing an entire building can be prohibitive, and even landlords can scarcely afford to borrow large sums at current interest rates. At the same time, the cost of such a project can double or triple if contractors are hired—whereas homeowners get free labor from themselves and their families and friends. Fortunately, apartment buildings often qualify for state and federal grants and low-cost energy-conservation loans which are not available to home-owners. But this is only part of the answer.

The trick to snugging up a large apartment building (or a small one, for that matter) is to forget for a moment whose name is on the deed, and not be intimidated by the scale of the undertaking. A twelve-story building still has only one basement, and there may be a lot of free help hiding out upstairs. Shivering. Weatherproofing is usually a do-it-yourself project, whether the street address is East 78th Street or East Peacham, Vermont. Simply start with the least expensive projects (which are generally the most cost-effective) and work up. Begin with stopping drafts, then look at the feasibility of insulating pipes, ducts, doorways, windows and the like. Savings from the initial work can help pay for more sophisticated weatherproofing measures later.

The following steps will usually pay for themselves in saved fuel within a few months, and then begin to show a net profit: sealing kitchen and bathroom vents, weather-stripping windows and doors, lowering water heater (and room) temperatures, tightening a drafty basement, installing tightly fitting window shades, insulating the water heater tank, installing heavy drapes (with a valance at the top to stop convection currents) and insulating heat pipes and ducts in the basement. Storm windows can pay for themselves within a year or two, depending on their cost. A furnace or boiler tune-up pays for itself in a few months. As in a house, insulation should go in the attic and basement first, and will save its cost in heating fuel probably within three or four years, even if installed by a contractor. If fuel prices rise, the payback periods shrink. And, of course, this does not have to be done all at once.

For a tenant, the best first move is to get together a few other tenants for a walk-through audit of the building, looking for obvious heat leaks. Get familiar with the structure and determine as precisely as possible what is needed, what it will cost, who can do the work and what the cost and comfort benefits are apt to be. There will be more than one weatherproofing alternative for every heat leak, so gather several options. The building superintendent or manager can be a help here, but be careful not to step on any toes. Winterizing projects are solutions, not problems, but owners and superintendents never like to be told that their building is falling down around their ears—and maintenance folks often consider the basement their personal domain. In any case, the more detailed your information is, the more persuasive your argument will be that something can and should be done. Be flexible, patient and ready to compromise, not only with the owner but with other tenants, who may not want to spend a Saturday insulating heat ducts. Finally, put any agreements you make in writing. Even among friends, memories fade when dollar signs are involved.

Remember, too, that sometimes all of this will work and sometimes it won't. People don't like disruptions, as a rule. They can be like that old skunk which lives under the tool shed and is forced to consider relocating when cold winter rains soak her bed. The place she is sleeping may be wet, but it's familiar—and God knows who you might bump into out in the barn.

10

Wood Gathering

≈ They say wood warms you twice, in the cutting
≈ and the burning. But it's a far trip from the woodlot to the hearth, and a winter's stock of cordwood
generally holds the power to warm you eight or nine
times before it sees a match. You must sharpen the
saw, then fell the trees and limb them and buck the
logs to length, then load them in the pickup truck, get
mired in mud, dig out, unload, split, stack, then walk
back down the lumber road looking for a pair of gloves
you left on the hood of the truck so you wouldn't forget them, all the while grateful it was such an easy day.
When woodcutting doesn't go smoothly it's often possible to stay warm well into December without ever
actually burning a thing.

Now, some claim there is enough pleasure in this
business that it might be outlawed or taxed if the government knew, and there is truth in that. But putting
up a winter's supply of firewood is still a good bit of
hard work, and buying it can be particularly painful, so
it's worth knowing how to get the most from this forest
fuel—from every bite of the chain saw and every log
that follows. Unfortunately, America's accumulated
store of wood-gathering wisdom, once passed down
carefully through generations, withered in the easy
arms of central heat—and a chain of knowledge reach-

ing back three hundred years was broken. The wood-burning skills of our grandparents are now being slowly relearned. There is an art, and some science, to the matter.

One of the first lessons of wood heat has to do with Roy Flowers. Since 1938, Roy Flowers has owned the Clark & Wilkins company at 128th Street and Park Avenue in New York City. Among other things, he sells firewood. Not the kind left strewn about my woodpile when April comes, but Sunday-go-to-meeting logs of oak, rock maple, birch and even apple, clean and straight and seasoned a year or two, with no loose bark or gnarled knots. His wood costs $8 to $15 a burlap-bag-ful. There are 60 bags to a cord. So a cord of Clark & Wilkins' wood at 1981 prices cost from $480 to $900. The economics of such a purchase are a little shaky.

As a *very* rough rule of thumb, one cord of seasoned hardwood burned in a good wood stove yields about the same usable heat as 180 gallons of fuel oil in a typical furnace. Which means that oil would have to be selling for about $2.70 a gallon for wood heat to make sense at $480 a cord. Clearly, the New York folks who call up Roy Flowers for a bag of firewood have something in mind besides economics.

We all do. A wood fire brings us back to a cycle of nature that we left long ago. It stirs a thousand musings of what was, or might have been, or could be. And like the sea, it is something you can watch for long intervals without giving explanations. A Maine newspaperman and author named John Cole once estimated that there are about a score of dreams in every half cord of wood. He was probably thinking of oak or maple. White birch yields a few extra dreams, and apple or cherry takes you to another land where money and cold don't exist at all. Even in October's silent woodpile there is a warmth, satisfaction, self-reliance that doesn't come with other fuels.

In effect, Clark & Wilkins sells about two hundred cords' worth of mental health every year.

From the practical standpoint of winter heat, however, there are an armful of other factors to consider in wood burning.

A cord of wood is a stack 4 feet high, 4 feet deep, and 8 feet long. It is a legal measurement, within reason. If you order a cord, no matter how long or thick the logs are, they should stack up to about 128 cubic feet. If you cut and split a full cord of thick, 4-foot logs into smaller pieces, they will restack into less than a cord because of the tighter packing and sawdust loss (which can amount to a surprising 10 to 15 percent of the wood volume). A standard pickup truck will carry half a cord of hardwood at best, if the wood is dry— which is worth remembering if it is delivered that way.

A "face cord" or "rick" is a vague term meaning a pile of wood usually 4 feet high and 8 feet long. It may be any width. When buying wood this way, be careful, and determine the length of the logs beforehand.

The heat value of a cord of wood relative to other fuels is virtually impossible to measure, because it depends upon the species of wood, its dryness, how efficiently it is burned, the efficiency of the fossil-fuel furnace it replaces and other factors. In general, a cord of dry hardwood can be worth about 180 gallons of heating oil, 28,000 cubic feet of natural gas, 5,000 kilowatt-hours of electricity or slightly less than a ton of anthracite coal. Expert estimates of these values are somewhat arbitrary, however, and vary as much as 30 percent. A good rule of thumb is that cordwood is generally still worth buying in rural areas as a heat source, but not in the city. At 1982 prices, the closest competitor of wood was coal (the two are about equal on a heat value per dollar basis), followed by natural gas, oil and electricity, in that order. Wood and coal prices, however, tend to vary more with locality than the others, and so are not always a bargain. Yet.

The potential heat value in a cord of wood depends on two things: the type of wood and how dry it is.

On a pound-for-pound basis, wood is wood. No species produces more heat than any other. But a cord of oak is nearly twice as *dense* as white pine, so a cord of it contains twice as much fuel. Hardwood generally (but not always) means deciduous trees, such as oak, maple and beech, which lose their broad leaves in the fall. These species are almost universally favored as firewood, because they burn slowly and evenly; produce hot, long-lasting coals and require little more effort to cut and split than softwoods (which are usually evergreen conifers, such as white pine, spruce, fir and the like).

Those species which produce the most heat per cord are shagbark hickory, black locust, white oak and American beech. Fruit trees such as cherry and apple are also superb, and give off a wonderful aroma when burned. Save them for special occasions; dead fruit trees are hard to find, and live ones have better things to do than make heat. Fresh-cut beech has a low moisture content, so it dries quickly when split, and burns readily if only partly seasoned.

The next best firewood trees include white ash, sugar (or rock) maple, yellow birch, red oak and pin oak. Ash wood contains even less moisture than beech. Yellow birch, like white birch, produces more sparks than most hardwoods, and can rot easily if left unsplit. Oak and sugar maple generally produce the best coals.

Red maple, white birch and American elm give off a

moderate amount of heat per cord. Elm can be virtually impossible to split because of its twisted, interlocking fibers, and someone once said, "it burns like churchyard mold." White birch burns relatively fast and produces a grand fragrance, and its bark will burn even when soaking wet.

The trees in these three categories all provide a high amount of heat per cord when compared to the softwoods, which tend to burn fast, spark heavily, and leave no hot coals worth talking about. They are lightweight, however, and make superb kindling for a stubborn fire. Sparks are less of a problem if the wood is thoroughly dried. Softwood is fine for a tight woodstove, if you don't mind reloading every half-hour or so, and are careful that the resulting hot fire doesn't overheat the stove or cause a chimney fire. Every wood has some value, and none should be left to rot.

Except perhaps poplar. The stuff sparks as wildly as spruce, gives off a bitter smoke, and burns like *wet* churchyard mold.

Unfortunately, it is difficult to tell many trees apart once cut into logs. Most softwoods can be identified by their light weight and pitch. Oak bark is generally deep fissured between craggy, fingerlike ridges. Red maple has more delicate, elongated ridges, and sugar maples have a smooth, almost flaky bark. Elm bark is even more flaky, and much softer, feeling almost like stiff cardboard. Beech has smooth, thin bark, and poplar has flat, scaly ridges.

In every case, wood burns more readily and efficiently when well seasoned. That means at least six months of drying under cover, and preferably a year. Most green wood contains about 40 to 50 percent water by weight. For example, a freshly cut, inch-thick maple plank, 6 inches wide and 12 feet long, holds an astonishing 5 to 6 gallons of water. The water content of most wood levels off as 20 percent if the wood is dried outside under cover for a year, and to about 10 percent if stored in a dry, warm part of the house. Well-seasoned wood not only burns more efficiently (the water doesn't have to be driven off by boiling), but will light with no more than two or three sheets of crumpled, twisted newspaper as kindling. Overdry wood may burn faster than you like, but loses nothing in fuel value unless it rots. Logs in which rot has penetrated the heartwood (the dark center of a log) can be discarded, because they will smolder, not burn.

The butt ends of a properly dried log will be gray, with hairline cracks radiating out from the center. Its bark may be falling off, and two logs knocked together will make a dry, sharp whack. Green logs make a dull thud, and the butt ends will be lighter, with no cracks.

When you buy wood from a dealer it is often hard to tell how well the wood has been dried. There is no legal definition of "seasoned hardwood," and at least some of the wood is apt to be green (or at least wet, if it was exposed to rain). Dry wood commands the highest prices, so the best bet is to order cordwood a year ahead of time, specifying green wood to get the lowest possible price, then dry it yourself.

To season properly, logs should be split and stacked out of the weather. I like to split anything more than about 6 inches in diameter. Stack the wood off of the ground to provide good air circulation and to keep the bottom logs from rotting. The simplest system is to lay two parallel lines of split logs and stack the rest of the wood atop that. (Heavy planks laid across bricks or cinder blocks also work well.) When building the woodpile, at the end of each row lay two or three logs perpendicular to the others, log cabin style. The ends of the pile can be built vertically that way, and the weight of each succeeding layer will keep the logs below it from rolling out.

Wood dries fastest in a sunny location, with the front of the stack facing south. It can be covered with canvas, roofing felt, polyethylene sheets or any other waterproof material. Polyethylene is popular, but it tears easily, is nonbiodegradable, and turns brittle when exposed to sunlight. In any case, don't drape anything over the woodpile to ground level, or moisture will be trapped inside.

For serious wood burning, the best solution is a permanent woodshed to keep rain, ice and snow crusts off the logs. It will also assure a good supply of dry bark and wood slivers for kindling, and provides a shelter for foul-weather log gathering and splitting. In the often snowbound North Country, a woodshed helps you remember in January where the firewood was stacked in October.

In most cases, about all that's required is a roof. If driving rain or snowdrifts are a problem, loose walls can be added, leaving ventilating cracks between the boards. Leave the front of the woodshed open for easy access, and because no wall can be built to look better than stacked wood. If needed, clear polyethylene can be fastened to the walls and across the front to create a sort of solar dryer. Leave air spaces at the top and bottom for ventilation. Green wood can be dried in three to four months this way.

When buying wood, it's often economical to order it in the spring, when demand is low. The longest logs are generally the cheapest. Fireplaces are suited to 2-foot lengths, but wood stoves usually need shorter pieces. Most dealers sell 4-foot logs, to be cut by the buyer, but 8-foot lengths are becoming common. In some areas it's even possible to buy these big logs directly off a flatbed lumber truck, several cords at a time.

Prices on all wood vary considerably, so call several dealers and try to look at their wood before ordering it. It can be cheaper to chip in with neighbors on a large quantity, rather than buying small loads individually. Find a good dealer and stick with that one. Good customers tend to get the best wood at the lowest prices.

Cutting and splitting large logs to stove or fireplace length goes more than twice as fast with two people than it does with one, but more than that can be dangerous. Stacking the woodpile is rewarding work that you are apt to want to do alone.

A good splitting maul is usually the only tool you will need for splitting wood. For large or stubborn logs, a pair of six- and eight-pound iron splitting wedges come in handy. One is usually enough, until it gets stuck in a green beech log. Then two will make you happy. A small axe is good for splitting kindling, but hatchets are dangerous anachronisms that can fly out of your hand easily, and serve no purpose that I've found. For splitting small billets 12 to 16 inches long, a splitting axe with a single-bladed head that has a wide bevel and weighs five or six pounds should do the trick. All of these cutting and splitting tools should be kept sharp with a file and whetstone.

The best time to split wood is on a subfreezing day when the wood has been aging for at least a few months and is frozen through. Winter does a sort of penance by making log splitting easier. Wrap heavy tape around the maul handle for the last five inches before it meets the head, as handles tend to splinter there. Set wood billets on a splitting block or wide stump to provide a solid base that won't dull the maul when it hits. It will also make the work somewhat easier on your back, by raising it above ground level.

Aim the maul to strike along a radial crack. In mid-swing, bend your knees so the maul penetrates at a 90-degree angle, and won't split your foot if it bounces off the log. Splitting is easiest if you swing *through* the wood—that is, if you pretend the log isn't there at all, and aim at its base. Some old-timers recommend twisting the maul slightly just before it hits, but this takes practice. In any case, cracks seem to open the quickest if a log is struck near its edge, rather than in the center. This also helps keep the maul handle from splintering.

Finally, wood should be split in the direction it grew (stump end up). If the log has a large knot or limb stub, split it so the knot will be cut in two. When that doesn't work, split a slab or two off the opposite side and then attack the knot section.

Never lay a log on the ground, straddle it and try to split it that way—unless your legs and back are made of stronger material than the maul.

For felling trees and cutting the wood to length, bow saws, two-person crosscut saws and even axes are still being used, and a sharp bow saw will always be one of the handiest tools ever made. But for serious wood gathering, say more than a cord a year, a chain saw is hard to beat. Today's models are lighter, safer and generally more dependable than ever. The old chain saw my family used twenty years ago weighs nearly twice as much as the one I have now, and it seems to start only on odd Tuesdays in April when there is a quarter-moon.

Chain saws with a short bar (12 inches or so) cut limbs fast and are cheaper than larger saws, but are not very powerful and have limited uses. For all-purpose woodcutting, a 16-inch bar is about right. An excellent safety device on new chain saws is a chain-brake, which stops the chain outright if it kicks back while cutting vertically into a felled log. Kickback is caused by the nose of the bar or the upper teeth hitting a hidden nail, piece of old barbed wire, tree limb or other object, and jerking the saw suddenly back toward the operator. It is the most common cause of chain saw accidents. A kickback guard, which protects the hand and forearm gripping the front handlebar, is another good safety item.

In selecting a saw, learn how easily the chain tension can be adjusted, and imagine trying to do it on a bitter day with numb fingers; on any saw, the chain tension will need to be adjusted regularly. Before buying one, compare several chain saws from reputable dealers who can also service them. The model you choose should feel comfortable and be relatively simple to operate and maintain. You'll be together a lot under adverse conditions.

Before buying a woodlot or ordering firewood from a dealer, check on potential local sources of free cordwood. Dumps and landfills often have usable wood to give away, as do local highway departments and utility companies, especially after a big storm or when a road is being built or widened. Housing contractors often know where a lot is being cleared for a new house. Supplies from such sources can be sporadic, however. Sawmills and lumberyards can provide a steadier source of waste wood, and state and federal forests often allow the public to gather dead or diseased wood for a minimal fee.

Partly because of wood burning, pickup trucks are now becoming a suburban vehicle. If you don't have one, or can't borrow one from a neighbor occasionally, have any cordwood you buy delivered to the house. It will be worth the time, labor and gas saved in making endless trips hauling wood in a station wagon or car trunk.

The art of felling trees and bucking them to firewood length is a serious matter that should be studied carefully, but is too involved to cover in detail here. There are several good books on these subjects. One of the most simple and straightforward guides I've seen was in the October, 1977, issue of Blair & Ketchum's *Country Journal*, Box 870, Manchester Center, Vermont 05255. It is available as a reprint for $1.

My own advice is to not only respect the chain saw and the tree it is felling, but fear them. Either can kill you. Wear a hard hat as protection from falling dead limbs (called "widowmakers"), and use heavy gloves. Work with another person, and leave yourself a clear exit from any tree you are cutting (and *use* it, even if the tree is falling exactly as you planned). Know that no two trees will ever behave quite the same way. Once a tree is down, it is still full of force. Watch for limbs bent under pressure, which can whip back quickly when cut partly through, and limb the tree so it won't roll over on you. Felling timber which gets hung-up in mid-fall is especially tricky to deal with, and should be treated as circumspectly as a bear trap.

There is warmth, but no nonsense, in woodcutting.

11

Wood Stoves

≈ Several years ago, the oil company noticed that
≈ our fuel consumption had plummeted, and sent
us a form letter saying that because we had begun
heating with a wood stove they were stopping regular
deliveries to our house, and we should call if we
needed them.

It was the ultimate compliment, but it wasn't true.
We weren't using the wood stove yet at all; we had
merely insulated and tightened the house, lowered
the thermostat and were, in fact, heating with old
eighteenth-century fireplaces which had been built in
the days when oil still came from whales.

Too often, wood stoves are assumed to be the first
line of defense in the battle to conserve energy and
stay warm on a lean budget. It shouldn't be. Wood
burning doesn't save heat; it substitutes one source of
energy for another, and our woodlands are no safer
from wasteful plunder than any other natural resource.
Europeans know that, and long ago devised some of
the most efficient wood stoves in the world to squeeze
every last BTU out of a piece of wood or dry peat. For
now, we are harvesting the most accessible wood—
trees that grew almost undisturbed for generations,
while we wallowed in a glut of fossil fuels. But soon we
may find ourselves cutting deeper into the forest, and

the price and availability of firewood will change.

Happily, a forest can renew itself every thirty years or so from a cordwood standpoint, and a well-managed 1 acre woodlot can yield a half-cord of firewood every year until the end of time. That renewable-energy aspect, and (until recently) low prices, helped spark the return to wood burning.

For a wood stove to be economical, however, you need (1) a steady supply of inexpensive fuel, (2) a house that can be adapted for space heating from one or two sources and (3) a willingness to tend the stove regularly so it operates at the most efficient level possible. Chilly mornings, bits of bark on the floor and callused hands also come with the territory.

As a rule, a wood stove will be only a will-o'-the-wisp heat source in a house with many small rooms, hallways and high ceilings. With vents, stairways or open lofts, convection currents from a wood stove can warm upstairs rooms relatively well, and unused rooms can be shut off entirely. But stove heat doesn't travel readily from one room to another. With high ceilings, it will pool-up well out of reach and just sit there, warming nobody. If the stove is used in conjunction with a central fossil-fuel furnace, and is located near a thermostat, parts of the house can freeze—and you may find yourself boosting the thermostat to astounding levels just to warm up out-of-they-way rooms, while using more fuel than before.

Wood stoves generally work best in large, open rooms where convection currents to other parts of the house are direct and simple. (See Chapter 7, Heat Flow.) The warmest part of the house will always be next to the stove. If your central heat works on a single zone, or there are many rooms on one level, trying to distribute heat evenly from a wood stove will be a finicky, frustrating business. If a stove is used to augment central heat, the house should have several zones, and the stove should be installed in a well-used common room, or keeping room.

Some people put a wood stove in the basement, where it won't affect the thermostats, and vent the heat up to the first floor. Unfortunately, much of the wood's energy is lost in the basement, and first-floor joists can interfere with heat flow. In some cases the system works quite well, however—it depends on the house—and a stove which can be fitted with warm air ducts is ideal in this case. The Garrison stove is one example, but there are others.

(Coal stoves behave the same as wood stoves from a heat flow standpoint, but have ash grates and are built to withstand much higher temperatures. So never burn coal in a wood stove unless the manufacturer's instructions say the stove is safe for coal, or can be adapted.)

A good wood stove is expensive, but a cheap one will cost more in the long run by wasting fuel, wearing out and being hard to load and adjust. So buy one that you can hand down to your grandchildren. They may need it more than you do.

On average, a good airtight wood stove will be from 40 to 60 percent efficient. That is, 40 to 60 percent of the fuel energy stored in a log will warm the room. The rest goes up the chimney. Leaky, nonairtight stoves are about half that efficient. They require constant reloading and tending.

The best airtight stoves have adjustable air inlet vents which, when closed, allow virtually no oxygen to reach the fire. When fully open, incoming air feeds directly into the heart of the fire, burning it hot and strong. With the vent (or vents) open a crack, a wood fire can be kept burning for six hours or more. (Twelve-hour burns are rarely found outside a laboratory.) Secondary combustion of exhaust gases is usually accomplished through a variety of baffles, downdrafts in the firebox, air passageways or secondary heat chambers, which either burn combustible smoke by channeling it back over the hot fire or capturing it in a smoke chamber to milk its residual heat before these exhaust gases leave the chimney. Such designs not only improve the stove's efficiency, they help reduce dangerous creosote formation in the flue.

Double-drum stoves, which have a secondary chamber to capture exhaust heat, are considered among the most efficient. Scandinavians still produce some of the finest wood stoves in the world. Beyond that, stove design is largely a matter of preference. Efficiency and performance depend mostly on how the stove is operated, not how it's designed.

There are a few other things to consider in choosing a new wood stove, though. First, a stove made of heavy cast iron or soapstone takes a while to heat a room after firing, but it produces steady heat and will stay warm long after the fire dies. A stove of lighter steel heats faster and cools quickly because it lacks thermal mass. Heavy stoves are best for living areas where overnight warmth is needed, whereas a steel stove is more suited for occasional use in, say, a workshop.

Enamel finish costs extra and is hard to find on many American stoves, but it holds its beauty indefinitely. Flat black finishes often rust, and can turn a dry, gray color after burning hot. Special paints and polishes are made for renewing the finish on these stoves.

More importantly, look for quality work. All parts

should fit tightly. Seams joined with furnace cement or by welding should look smooth and even. Spot welds can indicate air leaks and poor quality work. Legs, door hinges and latches and air inlet vents should all work smoothly and snugly. In buying your first wood stove, look at one or two models from manufacturers whose reputations are known and unquestioned; then compare the design and the quality of the work to other, similar-sized but cheaper stoves. A sticking air inlet could be nothing—or it could be a clue to generally poor quality and materials. Some new stoves are virtual clones of the finely tuned Scandinavian brands which have been around for decades, but are cheaply and often shoddily made. And that bothers me. Some of the most dependable brands include Jotul, Lange, Sevca, Riteway, Morso and Fisher. There are many others.

A common mistake is buying a stove too large for the area being heated. The result will be either a bake-oven environment with open windows for breathing, or low, smoldering fires, which not only burn less efficiently but add prodigious amounts of creosote to the stove flue. Manufacturers generally specify how many cubic feet of living area a stove will heat. If your house is well insulated and tight, you will want to be at the high end of that range.

To get the most from a wood stove and reduce creosote buildup at the same time, always burn dry wood which has been cut and split to various sizes and widths. A small, hot fire burns more efficiently than a big, smoldering one, and gives off the same amount of heat with less creosote. It is best to adjust heat output by varying the amount and size of wood you add to the fire, rather than with the air inlet controls. Experience will show how to do this.

Green wood doesn't exactly produce more creosote than dry wood, but heat that would otherwise go toward burning the wood and warming the room must instead be used to boil away the excess moisture—and that results in incomplete combustion and low flue temperatures, the two conditions necessary for creosote formation.

Reload the firebox on a bed of hot, glowing coals if possible, then open the air controls for a few minutes to get a hot fire again. All wood stoves smoke for a while after being reloaded, and these measures help cut that "dirty burn" time. When loading for an all-night burn, open the draft control fully for ten or fifteen minutes to heat the wood and drive off water then close the air supply down again for a long burn. In stoves that burn wood from front to back, like a cigar, you'll find it helps to pull a pile of coals to the front now and then.

With dry wood, newspaper and a few chunks of bark are about all that is necessary to start a fire. The manufacturer's instructions should give the procedures for starting and feeding a fire. Practices vary, depending on the stove's design and operating characteristics.

After an extended burn of six hours or more, it's wise to burn the stove with the air controls wide open for fifteen minutes or so. That will help dry off and clean out most of the creosote which accumulated during the time of air-starved burning. However, this chimney cleaning must be practiced *daily* or after every long burn, or creosote can accumulate to levels where a hot fire will ignite these flammable deposits and cause a chimney fire, which will scare the hell out of you and maybe burn the house down. A chimney fire sounds vaguely like a freight train running through the attic.

Safety, in fact, is the most critical part of wood burning.

For two generations, few Americans have had to shovel, stack, stoke or even light the fuel used to heat their homes. A simple thermostat setting pointed the furnace in the right direction, and it cruised through winter on automatic pilot. Our intimate relationship with creosote was allowed to atrophy, and the result was an epidemic of chimney fires which has hardly abated since the 1973 Arab oil embargo.

A wood stove cannot be aimed; it must be navigated. This means paying regular attention to creosote, air inlet openings, the dryness of the woodpile, condition of the stovepipe and a score of other matters which make a stove safe. A U.S. Department of Energy study found that three fourths of all stove-related house fires are caused by improper installation or use of a stove, not faulty equipment. Expensive new airtight stoves have been vented into clogged or leaky flues, empty attics and even closets.

The three most common mistakes are using an old, unsafe chimney; having poorly joined stovepipe sections which come apart in a chimney fire and putting the stove or pipe too close to flammable materials, such as furniture, kindling, walls or wet socks. In fact, a good chimney is more important, and often more costly, than the stove itself.

Defects in an old masonry chimney may not be evident until the house around it burns down. The common faults include broken or missing bricks, crumbling mortar and lack of an intact flue liner— which is particularly important in a flue only one brick thick. Such old flues used to be used for gas or coal appliances, and aren't necessarily safe against the heavy creosote of an airtight wood stove. During a chimney fire, heat or burning creosote (which has a way of building up in crevices) can breach the masonry and ignite nearby wood paneling or joists. Masonry chimneys can get extremely hot from a wood stove fire under normal conditions. A fatal 1972 house fire in Virginia resulted from nails, which had been driven into a masonry flue, overheating and igniting paneling they were fastened to.

An old chimney should be cleaned thoroughly and inspected inside and out by an expert before being considered for use with a wood stove. The best answer is usually a new chimney of either masonry with a tile liner, or prefabricated metal pipe. Masonry is more expensive unless you do it yourself, but it stores heat well and provides good protection against a chimney fire. If at all possible, put the chimney indoors. An outside masonry chimney drains heat from the house and causes creosote buildup on its cool surfaces. Metal chimneys have two or three layers of pipe separated by air space or insulation.

Wood stoves are generally hooked into a chimney flue with sections of single-wall stovepipe. With airtight stoves, a 4- to 6-foot run between the stove and main flue is ideal. Flue gases in the stovepipe transfer extra heat to the room and room temperatures won't be cool enough to cause creosote formation in the pipe or seriously reduce its draft. The pipe should not turn angles totaling more than 180 degrees before entering the vertical flue, or the draft can be poor. Each joint in the stovepipe sections should be fastened with three sheet-metal screws.

If the pipe is entering an old fireplace flue, it is best to insert the pipe in a hole cut above the fireplace, not into the chimney through the fireplace itself. Inspection of the flue for creosote deposits will be easier that way (shine a flashlight through the fireplace damper). As a rule, stoves should never be hooked into a flue used by a furnace, hot water heater or other such appliance.

Single-wall stovepipe should be kept at least 18 inches away from any unprotected combustible surfaces such as walls, ceilings, furniture or clothes being hung up to dry. The stove itself should be at least 3 feet from anything combustible, including the firewood box, according to standards set by the National Fire Protection Association. Local codes may differ slightly.

If a wall is protected by asbestos millboard or 28-gauge sheet metal (available from a lumberyard or hardware store), with at least a 1-inch air space between it and the wall, the stove can be installed within 18 inches of this surface, and the stovepipe within 12 inches. In any case, it's best to keep the stovepipe 18 inches from a ceiling. The common notion that a wall covered with a noncombustible material, without an insulating air space between, provides much fire protection is absolute nonsense. Conducted heat can still reach the wall. Wood exposed to temperatures of only 300 degrees can eventually ignite spontaneously—and can turn to charcoal in one to three years.

Wall shields should extend at least 18 inches from either side of the stove, and should be 2 feet wide behind the vertical stovepipe. For protection from radiant heat and sparks, the stove should rest on a noncombustible hearth which extends 18 inches away from the stove on all sides. This hearth should insulate combustible flooring against the stove's heat.

After installing a new stove, it's wise to check for creosote deposits twice a month at first, and then regularly after that. Clean the flue at least once a year. Every stove and every operator produces different amounts of creosote, and even a half-inch layer can be dangerous. Inspect the flue with a strong flashlight, either from the roof or by putting a small mirror in the cleanout door. Stovepipe sections must also be opened and cleaned. *Never* use flammable liquids to start a wood fire—the stove acts like the jacket on a firecracker and a small explosion may result.

If a chimney fire occurs, call the fire department and close the stove's air inlets and damper to starve the fire of oxygen. A panful of salt or baking soda can be dumped into the firebox; chemical extinguishers may also help—one should always be nearby, between the stove and an exit. Water poured on a hot stove can crack it and the flue. The best protection is a chimney that can contain a fire, and a wire screen on the top to keep burning embers from igniting the roof. There should be no leaks, cracks or faults anyplace along the installation.

Wood might warm you nine times, but creosote often heats a house just once.

12

The Open Hearth

≋ To the list of things you don't admit in public
these days, add one more: heating with an open
fire. After a brief respite as a reliable backup system
when rural power lines used to fail in a heavy dew,
the fireplace has become the official rumble seat of
wood burning. It is warmly regarded, but drafty,
and not much favored for steady heat against the
howling nights of a North Country winter. Worse, it
has a reputation for being inefficient. And now it's
more popular per capita on the West Coast than in
New England.

Say good-bye to the fireplace.

Or should you? A respectable wood stove will convert 40 to 65 percent of the energy contained in firewood into usable room heat (better results are seen
under controlled conditions and in manufacturers'
claims). Fireplaces, on the other hand, get a humble
zero to 15 percent efficiency rating from most wood
heat experts—and their *net* efficiency may actually be
negative if a wood fire is roaring in tandem with the
furnace on a cold day. In this case, the resulting chimney draft can drain more heat from a house than the
fire contributes.

Unfortunately, little quantitative study has been
done of fireplace efficiency, even though in New

England there are roughly three fireplaces for every wood stove, and about twenty million fireplaces nationwide. Until recently, performance estimates leaned on four tests done between 1823 and 1952, two of which used coal or gas, not wood, as the fuel. Only three studies have been published within the last ten years, and these involved prefabricated inserts or free-standing units. Several models were tested, but no comparative analysis has yet tried to find the most effective fireplace type, design or materials—especially in masonry construction.

Tests on a simulated masonry fireplace by the Shelton Wood Energy Research Laboratory in Santa Fe, New Mexico, in 1979 found the unit to be 20 percent efficient on the basis of radiant heat output alone, but no actual measurements have been made on the massive early-American stone or brick fireplaces and chimneys which were once designed entirely for heat. Not atmosphere or elegance—just heat.

Nor is data available on the heat-storing thermal mass capability of these interior structures. As an Auburn University wood-burning expert explains, "It's hard to get a . . . masonry chimney into the lab."

More importantly, wood energy research has paid little or no attention to how an open fire is built and tended. In a wood stove this can be important. In a fireplace, it is everything. Evidence suggests that the efficiency of a typical firebox can be doubled or tripled if it is worked properly. A slow-burning overnight fire of banked coals, while providing less heat, can probably outperform the best wood stoves.

Still, an efficiency range of minus 10 to plus 15 percent is far more common. The American fireplace has not been designed for heat output for at least a hundred years, since the age of central coal furnaces, and the ancient art of tending a smooth-burning hearth fire from November to May has long been forgotten.

Despite its limitations, an open fireplace does two things remarkably well:

1. Its ample air supply allows high firebox temperatures and near-complete combustion. Volatiles which could otherwise be lost up the flue, or condense as creosote, become a source of room heat. Chimney smoke is relatively clean compared to the air-choked, smoldering fires of a wood stove, where *combustion* efficiency (not overall efficiency) is low.

2. Fireplaces give off prodigious amounts of direct radiant energy—a highly efficient method of heat transfer through air. Unfortunately, radiant heat doesn't turn corners well, so one fireplace per room is the rule. Even then, one side of your body is apt to feel warmer than the other.

The most serious drawback of an open fire, however, is that it pulls in more air than is required for combustion, warms it and sends it up the chimney along with some furnace heat. The average fireplace draws two to five times as much air as a wood stove. Not only is heat lost up the flue, but extra cold air may be pulled in from outdoors to replace it. However, the cold air infiltration rate in an average house is greater than the draft rate of a wood stove *or* fireplace, which suggests that much of the cold air pulled inside by an open fire would be coming in anyway. The amount which can be blamed on the fire depends on the strength of the chimney draft. Keep it as gentle as possible without smoking up the room.

The best way to keep a fireplace from draining away furnace heat is to keep the furnace turned down. Fireplaces work most efficiently in cool houses. They won't heat the next room well, except by residual chimney heat, but they provide a focus of warmth and life that no other source of winter heat can match.

Much of the air drawn by an open fire serves no purpose but to carry off smoke. Therefore, the most energy-efficient fireplaces are designed and operated to produce maximum heat output while discharging smoke with the quietest possible draft. The firebox and chimney should also have the capacity to conduct into living areas heat which would otherwise be lost up the flue, and the house should be as tight as possible to control cold air infiltration.

In one sense, a house with fireplaces can be seen as a sort of giant wood stove where the family lives inside next to the fire. The air inlet is leaks in the house envelope. Some people try to localize and regulate incoming air with an insulated duct which connects the outdoors with an adjustable vent near the firebox. My own feeling is that this isn't worth it; a fireplace draft is hard enough to fine-tune without worrying about a vent—which is apt to be left open more than necessary anyway.

The efficiency of a fireplace depends greatly on its design. The best is the so-called Rumford fireplace, which became popular in England in the late 1700s and is being copied again by a handful of masons. The Rumford—named after Benjamin Franklin's contemporary, the inventor Benjamin Thompson (Count Rumford)—is a refinement of early colonial designs. It has an unusually shallow firebox and high lintel to bring the fire almost into a room, with sharply angled rear and side walls (called the "fireback" and "covings") to reflect outward as much radiant heat as possible. An unusually narrow flue throat and damper, often as wide as the firebox, are located as high as a

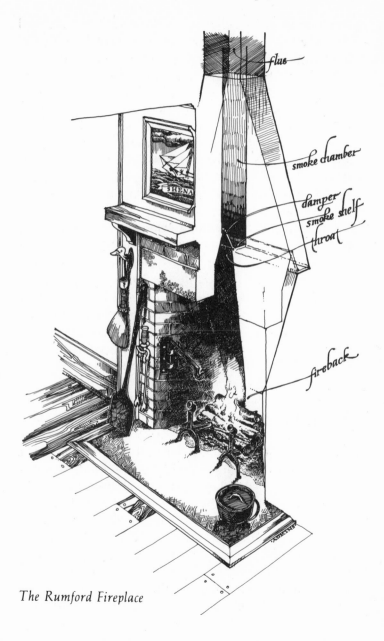

flue

smoke chamber

damper
smoke shelf
throat

fireback

The Rumford Fireplace

foot above the lintel, leading into a smooth-walled smoke chamber which looks like an upside-down, gently tapered funnel. A thin ledge, or smoke shelf, deflects sudden downdrafts in the slightly oversized flues characteristic of early-American chimneys.

If the system is built properly, smoke rises in a gentle, determined draw. I've seen some old Rumford-style fireplaces where cold air actually rolls down the back or one wall of the flue, off the smoke shelf and into the fire—while smoke moves lazily up the other side in an efficient, circular flow. One of these happens to be in our keeping room.

The Rumford looks like a dent in a brick wall compared to the low, cavernous, boxlike design of a typical contemporary fireplace, which reflects little radiant heat into a room. Subtle design flaws, such as a flue throat too low near the lintel or a squat smoke chamber (or none at all), often require a strong draft to pull smoke up the chimney. And that costs heat.

Some fireplaces are open on all or several sides, looking in the extreme like a funnel over a campfire. They expose flame to a wide area, but tend to reflect the far more intense heat of burning coals directly upward. Moreover, smoke likes to hug a wall, and open designs often require a strong draft. Prefabricated metal fireplaces, many of which are insulated to stand directly on a floor or fit into a wall, can have the same radiant energy output as a standard masonry fireplace, but they lack the power of masonry to conduct and store heat.

More than any other heating device, a fireplace is part of its chimney; and an interior masonry chimney stays warm long after the fire dies, acting as a massive keel to level out overnight temperature dips. Its ability to conduct heat also helps slow the often-powerful drafts resulting from hot, open fires.

Convective-type prefabricated units, such as the well-known Heatilator (often called "circulating fire-places"), throw extra heat into a room through air ducts located behind the firebox walls. Cool air is drawn in at floor level, warmed and sent back to the room through vents above the fireplace, either by natural convection or by blowers. Some models can be inserted into existing masonry fireplaces, but they are more common in new construction. These units can be up to 40 percent efficient if there are glass doors and air inlets that can be shut tightly against air loss, according to one study. The doors stop virtually all radiant heat, however, and the unit behaves somewhat like a low-efficiency stove. In addition, the air outlets are well above floor level, resulting in a problem shared by raised hearths: cold feet.

Every fireplace must have an adjustable damper that leaks no air when it's closed. The type installed on a chimney cap and operated by hanging chains is probably ideal (until it sticks) because it keeps cold air from filling the flue when the fire dies. When a fire is burning, dampers should be shut as far as possible without interrupting the smoke's flow, and they must be adjusted regularly—to fit changes in the fire and the weather. Drafts tend to be strongest on cold, dry days.

But all of this combined won't necessarily make the system much more efficient. Unlike a wood stove, even the best-designed fireplace can't do much on its own. Here, efficiency depends on how the fire is fueled and built and tended and closed down for the night. There is art, and a little magic, behind that.

The first step is to build up a good bed of ashes in the firebox. A *good* bed. One that's a foot deep in back, or more, sloping down to the front feet of the andirons. This insulating ash bed, common in a blacksmith's forge, lifts and focuses the heart of the fire into the logs—resulting in a hotter bed of coals and more complete combustion—while at the same time directing heat into the room. It will take a full winter to accumulate this much ash, even if a fire is burned regularly. And it should be burned regularly; that's the next step. A masonry chimney works most efficiently if it is warmed once in the fall and not allowed to cool off again completely until spring. This not only keeps the flue primed for a good, gentle draw, but reduces the creosote buildup which occurs regularly on a cool surface.

The fire itself should be almost a subtle thing: quiet and low, with little flame. But the bed of hot coals underneath should turn iron so red you can bend it, and logs should be arranged so that these coals are directly exposed to the room.

Always use dry wood. Green logs smolder, producing smoke but little heat, and the damper will have to be wide open to pull up this smoke. Dense hardwoods such as hickory, oak or sugar maple are far more important here than in a wood stove, because they produce the best hot coals with the fewest sparks. Popping sparks in an open fire require constant use of a fire screen, and that can cut radiant heat output nearly in half.

To build the fire, place a large log in back, up against the sloping ash bed where no flames will leak up behind it. Newspaper is generally used to start a wood fire, but it leaves a feathery ash which can float all over the room when you sweep up the hearth. I prefer birch bark, and peel it off dry logs for just this purpose. It lights with a match even when wet, burns intensely, smells wonderful, and leaves a good ash. Cardboard egg cartons, waxed milk containers and butter wrappers (or any waxed paper) also work well. If newspaper is used, crumple it and twist it in the middle for a slower burn.

Dry bark is good kindling and is plentiful in a seasoned woodpile, but any slivers of dry wood will do, and lumberyards often have scraps of dry wood to give away. I avoid pieces of boards with old lead paint on them, because the wood ashes end up in the vegetable garden. If the fumes aren't toxic, they are not healthy, either. Save old candle stubs to help start a stubborn fire. Rest a small piece between two logs to drip down on the burning kindling. This will start almost any fire, but don't use much—candle wax is highly combustible.

I generally set two small logs atop the kindling, in front of the big rear log, and fire things up. If the ash bed is high, you might want to burn the kindling for a minute or two to get things started before putting these logs on. Keep a space between each log so flames can burn up between them. With small, round logs, overlap the butt ends on one side, fan-style, to space the wood out. Start with small logs, maybe the size of your forearm, and work up as a bed of coals develops. An odd number of logs, three or five, works best. But there is no law on that—it depends on the size of the wood.

The idea is not to produce flame, but to work up a hot bed of red-white coals and firebrands in the bed of the fire. As a log burns through, push its ends toward the center so that the bed of coals is exposed to the room as much as possible. Pull the entire affair as far toward the front of the firebox as you can get it without smoking.

At some point in the fire's progress, it will reach

peak heat output. There is no way to describe this—you will simply *know* it. Coals will be deep and hot, with little flame from the wood. Seen from outside, the chimney will show no smoke at all. That's when you add a new log—just before the fire gets hungry. Add this log to the back, where it will cook quietly awhile, losing moisture. As it crumbles into coals in time it will fall forward, and it's almost time for another log. That should take an hour or two, but the fire will probably need tending every fifteen to twenty minutes or so. There is nothing easy about this fireplace business.

Once the fire is well established, it should need only one big log in back, with smaller chunks of half-burned firebrands in front. Always keep its skirts lifted so the coals are exposed to the room. Try to use the fire screen only when you leave the room. *Every* fireplace should have a fire screen.

In a properly running hearth fire, heat will be more intense 2 feet in front of the fire than above it because wood ashes insulate as well as fiberglass, pushing heat out under the logs and into the room. The fireback, covings (the two vertical sidewalls of the firebox), logs and ash bed are all designed to push heat out, not up. Such a fire, with its heart of intensely hot coals and little smoke, heats a room far better than a violently roaring and spitting blaze—and usually requires that the damper be open only a crack. Always keep the damper set as narrow as possible, without causing smoke to leak into the room. This takes practice and regular attention, but saves heat. Fireplaces work best in the long run when somebody is home all day to tend them.

Plan the fire so that when you are ready to close it down for the night, all that remains is a hot bed of coals and firebrands. Pile these up in the center of the ash bed and cover them with a deep bed of ashes until not even a small wisp of smoke appears. The coals will burn invisibly all night, and sometimes for a day or more.

This slow, "banked" fire requires little oxygen, produces virtually no carbon dioxide and needs no draft at all. The combustion is virtually perfect. Some people close the damper entirely and end up with a fire which is probably well over 90 percent efficient (all dampers leak slightly). At daybreak, the hearth is still warm on your feet, and ashes can be scraped off the still-burning coals for a new fire that requires no kindling or paper.

About the only tools necessary for fire tending are a good set of tongs which won't pinch your fingers when used with one hand, a shovel for banking the fire at night and scraping off ashes in the morning and a small hearth broom. I've never found the need for bellows or a poker. A long-handled "peel," once used for shoveling hot coals in and out of a bake oven, works perfectly for banking the fire at night, but a smaller shovel is better in the morning.

Many devices being marketed as fireplace efficiency aids are, in effect, trying to make a machine out of a hand tool. Some work well, though, and there are a few guidelines for choosing among them: (1) The accessory should be able to withstand high temperatures. (2) BTU ratings have nothing to do with efficiency. A house fire has a high BTU output. (3) If your major goal is efficiency, and the device you are considering costs nearly as much as a wood stove, buy a wood stove.

Glass doors are a poor choice with masonry fireplaces. They block radiant heat, and usually aren't tight enough to stop significant air loss.

Hollow metal grates which throw hot air into a room have been found in independent tests to be effective if they have a blower, but not if they don't. The blowers can be noisy enough to take the pleasure out of a hearth fire. In any case, grates tend to hold all but the smallest bits of wood well above the ash bed, so it's difficult to get a bed of hot coals burning under the logs. Andirons work best, and the horizontal arm should be no more than about 3 to 4 inches above the hearth.

Interior masonry chimneys are far more efficient than those built on an outside wall, because they conduct and store residual heat from the fireplace draft inside the house, where it's needed. An exterior chimney conducts this heat outside. That means the chimney flue will be cool, which requires a strong fireplace draft and leads to creosote condensation.

If your chimney is on an outside wall, there are several possible things to do. These include:

1. Build an addition around the chimney base to bring this extra heat inside the house. If the chimney is on a south wall, this is a good spot for a solar greenhouse, because the masonry can serve as thermal mass to store the sun's radiant heat.
2. Brick over the fireplace entirely and install a wood stove, using that flue. This is work for a mason, because the damper will have to be removed and a flue liner dropped into the fireplace, with a cleanout door at the bottom.
3. If the fireplace is of the deep, cavernous type, it may be possible to build a new brick fireback and covings in front of the old ones, to create a shallow, Rumford-style firebox. Insulate be-

hind the new courses of brick with vermiculite.
An experienced mason will be required here.

In the right hands, and the right house, a good fireplace can probably run as efficiently as a wood stove.
But most don't. Too much has been forgotten, and too
much can go wrong at too many turns. A low lintel, an
outside chimney, an hour's inattention or even too big
a log can all chip away at the net operating efficiency of
an open-hearth fire.

Or maybe that isn't the point at all. Could it be that
the thing an open fire does best cannot be measured?
What, after all, is the net efficiency of a Northern Spy
apple, or a county fair or the old setter who has slept in
the middle of the kitchen floor since as long as you can
remember? Maybe what a fireplace does best is not to
be a stove, or furnace or any other sort of heat machine. Maybe an expert is nothing but a damn fool a
long ways from home, as Carl Sandburg said. Or
maybe it's the expert who is right in this case, but not a
poet.

I don't know. I do know that I've tended fires on
hearths that were tended for centuries before. I've
smelled the same woodsmoke my father smelled, and
his father before that; and cold as the temperature on
my back may be, if there is warmth in this world, it is
in the glowing embers that can be wheedled and
nursed and called into fire as if some Merlin of the
hearth were in the wood itself.

And if all that is a mistake of net efficiency, it isn't
one of very great moment.

13

A Place to Build

≈ Somebody has calculated that the average American family moves into a new house every four to five years. Now, if you figure it takes a year to unpack boxes (and it does); a year or two to patch the roof, paint, wallpaper, fix the septic system and scrape dried food off the kitchen ceiling; then another year to repaint, repack and get the house ready to sell, it's a small miracle that anybody has time to snug a place up for winter. Most houses, I suspect, are probably drafty just from all the realtors filing in and out.

At some point, though, we begin to toy with the fine and sturdy notion that the next move will be the last. That here the packing ends, and the roots go down. The expectation, the *right*, almost, to own a place of our own may be uniquely American, but the sometimes more powerful need to settle somewhere permanently, or maybe just return to a place we know, where the soil and the smells and the faces are all familiar, is universal.

I remember once traveling in Russia, where love of the mother country (but not necessarily the mother government) runs as strong and deep as anywhere in the world. At one point, in Moscow, my wife remarked in an offhand way to the highly educated and wonderful young woman who was our Intourist guide that we raised a small herd of dairy goats. Marina, the guide, was astonished and clearly thought we were joking. How, she finally asked, could anyone raise goats in an apartment? The idea of somebody living in what she called a "separate house" was so totally foreign that it didn't register at all. Which, in turn, left us feeling somewhat humbled.

Of course, inflation and high mortgage interest rates are making "separate houses" something of a foreign notion here, too. The difficulty is not only in being able to afford to buy or build a place, but in holding on to it—in insulating it against every possible annual expense but taxes. This means, more than anything else, letting nature provide the heat. And the first secret to this lies in the land itself. Property is still affordable these days, even if houses aren't, but you may have to travel up-country a ways to find it. For now, let the house wait.

In northern climates the most practical house site from a winter warmth and energy-efficient standpoint is a south-facing slope with well-drained sandy loam soil, easy access to water and trees that break prevailing winter winds but allow sunlight and summer breezes to reach the house. Spend some time looking for this land, and follow it through the seasons if possible before deciding exactly where the house will go.

A topographic map, available in many stationery and sporting-goods stores, can save hours of trudging through swamps with a realtor in tow. In developed areas, try locating a natural scenic view looking north (popular lakes, mountains and so on), then walk over the hill for a "less desirable" and cheaper lot facing the other way. In New England, lilac blooms and oil truck's last winter visit can both come literally a month earlier on the south slope of a hill than on the north side, because of the sun's direct winter rays.

Sandy or slightly gravelly soil provides the best digging and drainage for septic systems, foundations and vegetable gardens (although loam is also needed here). It also provides a firm, natural base for access roads or a driveway. Shallow bedrock is fine for some foundations (not full cellars), but can result in costly site development for septic tanks. Clay soil causes heavy stress on cellar walls, foundation slippage and poor drainage. Check local codes for soil absorption requirements, and dig a hole several feet deep to check the soil firsthand. Then, if there is any doubt, have a professional soil survey done.

For natural vegetative shelter, the height, density and location of trees can affect your winter fuel bill and summer comfort. Deciduous trees on the south and west exposures will shade the house in summer without blocking the winter sun. A full stand of evergreens to the north or northwest (or the direction of prevailing winter winds) helps break the heat-robbing force of cold winds.

Even a tight house will lose heat in a 20-mile-per-hour wind at twice the rate it would in a 5-mile-per-hour breeze. A good shelter belt of evergreen trees and shrubs can cut most of this heat loss. Maximum wind protection behind a hedgerow occurs at a distance away of about five times the height of the hedge, so consider this in deciding where to plant trees and shrubs, and what type to use. It is wise to vary heights between low shrubs and tall ones, to get a full wall of wind protection. Because a belt of evergreens is relatively porous and able to bend, it is less apt to accelerate winds the way a city building does. The bare branches of deciduous trees provide some wind protection, so let the evergreens develop a bit before cutting these down—unless their shade is too heavy for the new plantings. Meanwhile, use the leaves from these and other trees to build a leafy berm of insulation around your foundation in the fall. This is more attractive than you might think, and by the time cold weather departs the leaves will have settled into a nice humus for the garden. Low shrubs, or even shoveled snow, also provide good winter protection around the house foundation.

As to the house, you might just consider building that yourself. But don't consider it lightly.

The enormous, almost defiant notion of building your own place can sneak in the back door, settle underfoot, tug, itch, then grow into a full-fledged decision while your back is turned. It can be fueled and stoked by exasperation over the numbing sameness of subdivision homes; by the longing to reclaim and know a thing abandoned to experts generations back; or even by the recollection of a proud, rickety sanctuary nailed together with your own hands one summer in the old maple tree beyond the wall, when life was simple and you could do anything.

It is a dogged notion, which can survive for months on little more than the knowledge that an average new house will cost nearly $100,000 in a few years.

Many who tackle the construction of their own homes are returning to the prenineteenth-century system of doing virtually all the work themselves, and hiring out only the most specialized jobs. At the same time, an increasing number are cutting some time and uncertainty ("fear" may be a better word) out of this process with a precut house kit. The two ideas are not unrelated, as most early New England homes had post-and-beam frames cut and fitted by a local joiner, then marked with Roman numerals and erected by the numbers.

There are three things to know about building your own house, whether from scratch or a kit. First, it is an enormous undertaking that can scare hell out of you and have a similar effect on neighbors and banks. Second, it takes time—from months to several years, depending on your vacation schedule, the complexity of the house, the amount of work you do yourself and your definition of the word "finished." Third, and most important, it can be done—by almost anyone, and for substantially less than the cost of conventional construction.

"You need common sense, but patience can overcome a lack of skill," says Pat Weisel of Underhill Center, Vermont, who had never swung a hammer before becoming a full-time framer, carpenter and roofer on a reproduction antique Cape-style house. Her husband, Mike, helped nights and weekends.

All other things being equal, the cost of an owner-built house can be anywhere from 10 to 50 percent less than that of a conventional dwelling. Sometimes the savings are even greater.

The trick is to start small. Rambling old farmhouses were often built in stages, and can be again. Begin with a keeping room, kitchen and bath. Pile in insulation until the cows come home and take full advantage of solar heat. Build it tight, strong and comfortable, with an inside masonry chimney for stove heat. Whatever crafting ability you put into this snug dwelling will carry through into the rest of the house, and if fuel ever becomes too scarce you can always withdraw here after the main house is completed. This section of the house will be fully insulated and self-reliant, so it will always make a perfect winter keeping-room-suite.

One guideline to building your own house at a low cost is to learn how to do everything that is labor-intensive or that may have to be done more than once. For example, full cellars and septic systems require a bulldozer or backhoe, and are worth contracting out; whereas making doors, window casings and even concrete foundation columns requires lots of labor but little special equipment, and gets progressively easier after the first one.

Plumbing is a skill that can be learned fairly easily once you study the principles, understand the materials available and practice soldering a sweat joint on copper tubing. This skill comes in handy anyway when repairing frozen pipes or installing solar water heaters. Electricity is also relatively straightforward once you get beyond the service box—but in this case mistakes don't simply drip on the floor. They can kill you or burn down the house. In either case, codes usually govern how the work is done, and it's smart to work closely with local plumbing and electrical inspectors. They can be a big source of help.

Consider using a post-and-beam frame in the place. This early historical method of framing a house requires a high level of skill to cut the parts; but once cut, the timbers can be raised and fitted together in a day, and then roofed to keep the rain off your head while you complete the house. Post-and-beam frames have a relatively small number of large timbers, ranging in size from 4-by-4 braces to massive 8-by-12 girts and summer beams, jointed and pegged together and braced at the vertical angles. The system faded in the mid-nineteenth century with the advent of modern housing pressures and machine-made nails, but is finding new popularity because of its strength, flexibility, and visual appeal.

Exposed rafters and beams become a part of the house furnishings, and the frame allows flexibility by having no supporting interior walls and by rarely having many load-bearing vertical posts (or studs) in the outside walls. The open interior lends itself to stove heat, and the wall structure allows open window expanses on the southern exposure and permits insulation to be installed uninterrupted by studs.

Although you can learn most of the skills it takes to put up a new, heat-efficient house, this does not mean you will avoid mistakes and probably take two or three times as long as an experienced carpenter, plumber or electrician. You will. That's where the time goes. The knowledge can be gained from books (there is a bumper crop of guides to house building on the market), from helping somebody else with their house, or from one of several schools set up to teach the art of house building. Two of these, both of which have excellent reputations, are the Shelter Institute in Bath, Maine, and the Cornerstones School for Energy Efficient Housing in Brunswick, Maine.

Old-timers can be invaluable, but they are in short supply these days, and the more out of the ordinary your house is, the more ingenuity and research will be needed to build it. Few old Vermont natives, for example, know how to put a sod roof on a yurt. So stick to the simple ways.

14

The Body

≈ In the oral history of those who inhabit the barren
≈ Arctic regions north of Hudson Bay, the story is
still told of how Polar Eskimos once survived the win-
ters in western Greenland. When cold and darkness of
the long night settled across this frozen land, and winter caches of food dwindled, these people went into
their stone and peat houses, covered themselves with
layers of animal skins, and lay still. Week after week
they remained inside, using up fat and hardly moving,
until the sun's rising arc began to brush the horizon
again, and there was enough light to hunt for seals by
the blowholes that these ocean mammals keep open in
the ice all winter.

Hardship, it seems, is a relative thing—and this
small saga speaks as much for plain tenacity as it does
for the remarkable ability of the human body to adapt
and survive. I favor my own body enough not to put it
through such a test, but the fact that we might just pass
it is nice to know.

Whatever other odd pursuits our bodies may get involved in, they do a superb job of producing and stor-
ing the right amount of heat needed to stay alive.
Warmth is created by metabolism—the cellular act of
converting food to energy—which is regulated by the
small hypothalamus gland, located in the brain behind

the lower forehead. At rest, a body radiates as much heat as a 100-watt light bulb. While working or exercising, it will heat a room nearly as well as a large electric space heater. This metabolic warmth is distributed through the body primarily via the circulatory system, which is why smokers, drinkers, elderly persons and those with respiratory and circulatory ailments tend to get cold hands and feet sooner than others.

When a human body is subjected to cold, it begins to react immediately. First, the hypothalamus restricts blood supply to what it regards as relatively unimportant parts—the skin, hands, feet, ears and nose. These are the first places to get chilled on a cold winter day. Small arteries and capillaries in these regions constrict automatically, and the body's warmth is concentrated on life-sustaining organs such as the heart, lungs, liver, kidneys and brain. The head, in fact, is the warmest exposed part of the human body, and on a freezing day more than half of our body heat loss occurs in the head and neck. That is why it is important to keep those areas covered—and why they so seldom feel cold even without a hat and scarf. The face is protected by fat and a good blood supply, and Eskimo faces tend to be especially chubby.

(Eskimos, incidentally, are relatively short and squat compared to people from warmer climates. There are two theories for this. One is that most of their body energy goes toward producing heat, not growth. The other is that such a squat stature enables them to have the largest body volume possible relative to exposed surface area. Low, square houses are energy-efficient for the same reason.)

Meanwhile, as the body's blood supply to extremities is closing down as a defense against the cold, its metabolic rate is rising. In muscle tissue, this is brought about partly by the act of shivering—which is nothing more than a form of involuntary exercise. Shivering usually stops immediately if you begin to run or do other work. Metabolism also increases in organs and tissues which have no muscle, such as the liver, as a way of keeping these vital parts warm.

At the same time, goosebumps begin to appear, until parts of the body look like a plucked goose. The term is more accurate than you might think, because goosebumps are caused by the contraction of a small muscle at the base of every large hair follicle, and is probably a vestigial attempt by the body to fluff up its fur and provide more dead air spaces as insulation against the cold. Birds (such as geese) and animals still do this, and can look unusually fat on a cold day. For better or worse, humans have traded in their fur for clothes.

The best way to increase your metabolic rate and feel warmer on a cold day is to exercise. That can mean anything from stomping your feet to chopping wood, dancing or whatever. It is possible to get some sort of exercise almost anywhere. Eating food also raises the metabolic rate—not only by providing the body with fuel, but by stimulating the hypothalamus for a short period. With training, metabolism can even be controlled with biofeedback methods, and some Far Eastern monks have developed the ability to localize body warmth to such an extent that they can stand on ice barefoot for long periods without suffering from frostbite. (Frostbite occurs when the blood supply to an extremity has been restricted so effectively that the tissue begins to die.)

Hands and feet are designed to function well at near-freezing temperatures, however, and average hands have a skin temperature of only 92.4 degrees F.

Because of superior blood flow to their extremities, an Eskimo's hands are able to stay warmer in icy water than yours or mine.

Penguins, and a few other Arctic mammals, have developed an even better system. Penguins are able to walk on ice in pure comfort because of a sort of circulatory "heat exchanger" in their feet. Below the calf, their arteries divide into channels, with returning veins wrapped around them closely. Before the cold blood from their feet returns to the vital organs and signals the brain to close off the circulatory system down there, this cold blood has already been warmed by the arteries passing close by the cooler veins. The feet are chilly, but the body stays warm.

One of the worst enemies of the human body's natural inclination to stay warm and alive is alcohol. This deadens the hypothalamus connections and puts the body's metabolic rate in a sort of holding pattern which adjusts poorly to the cold. In fact, more than half of the people brought into hospital emergency rooms suffering ·from hypothermia—or severely lowered body temperature—are found to have been drinking heavily beforehand. Healthy people are rarely affected by this deadly loss of body heat, and then only in extreme circumstances.

Fat is the body's natural insulation, and women tend to have somewhat more than men—although men generally have a higher base metabolic rate. Fat insulates well because it has a poor blood supply, so is less subject to freezing than muscle tissue (note which part of a roast thaws quickest when taken from the freezer). Fat also has large, elongated molecules which contain little water—and water is an excellent conductor of heat or cold. So-called white fat serves mostly as insulation for the body in general, and for particular organs. "Brown fat" not only insulates but produces heat, and is found in heavy concentrations in newborn babies and in Arctic animals.

The effect of cool room temperatures on health (assuming the body itself is clothed and warm) has not been fully determined, although specialists consider cold air and high humidity healthy for the skin. There is, for example, no medical evidence that bare heads or wet feet cause colds. One recent study indicated that some viruses and bacteria do better under chilly conditions, but most winter colds are probably spread because people tend to gather close together indoors between November and March. While some cold germs thrive at 55 degrees, others undoubtedly do best at 75. At the same time, some doctors recommend that patients with winter respiratory ailments, such as dry throats and sinus conditions, sleep under warm blankets in a cool (50- or 55-degree) room, rather than use a humidifier. Low temperatures stimulate the mucous membranes, whereas warm indoor air tends to dry them, unless the house is unusually humid.

Many people, myself included, find they have fewer colds and generally feel healthier when indoor winter temperatures are kept far below what heating engineers would consider the "comfort zone." Our preschool daughter, whose room seldom gets above 60 degrees in winter, has not had a cold in two years—although medical science wouldn't draw too many conclusions from that.

It *has* been shown, however, that it takes only a week or so of steady exposure to cold for a healthy human body to adjust its metabolism and feel comfortable again. This is one reason why a 40-degree day in October sends a raw chill into your bones, while the same temperature in January would be absolutely balmy.

And many who live in cool homes can hardly bear the stifling 68-degree air in stores and restaurants.

Not many years ago, at the University of Oslo (Norway), a group of student volunteers slept outside in subfreezing winter cold for a week with little more than a cot and blanket. For the first few nights, they were miserable. But their bodies adjusted until, after a week, they hardly noticed the cold. Then, of course, they found it virtually impossible to sleep back in the dormitory again without an open window.

Elderly folks and infants can be special cases in much of this, however. Older persons often have less efficient circulatory systems or other ailments which make them more subject to the cold, and may be less apt to boost their metabolic rates with activity. Every human body is different, though, so don't assume that everyone over seventy should stay wrapped in a blanket all day. Too often, older people can begin to see themselves as delicate—or even senile—because we treat them that way and expect it of them. I remember buying some rough-cut boards from an old water-powered sawmill in New Hampshire on a winter day some years ago, and being somewhat startled to find that the owner, who was in his sixties, did most of the lumber-cutting himself. Until he mentioned that his father helped him.

Newborn infants are another matter, and require special attention. The hypothalamus and metabolic system is not fully developed until about two weeks after birth, longer in premature babies. Until then, rooms should be kept at about 68 degrees to prevent chilling. The body of a newborn infant can lose heat quickly because of its relatively large surface area and small volume. And newborns have scant reserves of fat and food to draw on, and move around relatively little at first. So keep them warm. A chill is easy to detect because babies know how to shiver and get goose-bumps. It's a good idea to warm an infant who actually feels chilly to the touch, before wrapping the baby in layers of blankets, which may simply hold in the cold. Holding the baby close against your own skin is about the quickest way to do this.

Within a few weeks, however, babies quickly become survival machines by building up their body weight at a rate almost unequaled at any other time of life. Once their systems are regulated, they accumulate fat quickly. Children also have high metabolic rates because they are so active, and can stand cold better than adults. Youngsters who are able to talk rarely complain of being cold before their parents feel chilly (unless they are sick), and more than one parent has been startled to see something that looks like a frozen snowball on feet running inside on a December day to look for dry mittens, shedding hats and coats all over the kitchen floor, only to disappear back outside again.

Mention hibernating in western Greenland to a nine-year-old, and he probably couldn't wait to get going.

15

Clothes

≋ Bonwit Teller, that purebred bastion of feminine style and elegance, advertised wool long johns in its fall clothing catalogue last year. Imagine that! Woolies—right there at the top of the page with properly daring evening gowns, lingerie and all manner of things you hear about but never actually see women wearing. The Bonwit Teller version, of course, looked nothing like what my Uncle Henry wore on his farm in winter (for one thing, Uncle Henry wore clothes over his, and not everybody is doing that these days), but they serve the same purpose in the end, and that is comforting to know.

Good clothing is the most effective insulation there is because it keeps warm the only thing in a house that really cares—without wasting any effort on tables, chairs, silverware and the like. And the first thing to know about good clothing is which materials will keep you warm and which won't.

Wool will. Virtually no other natural or synthetic fabric has wool's natural insulating ability, resilience and general quality of appearance. It is easy to find, easy to work with and tailor, is flame-retardant and does not stain easily—especially if the natural lanolin oils have not been processed out. These oils have a pleasant, but somewhat earthy, odor, and serve the additional function of helping wool repel water. Unlike other fabrics, such as cotton, wool does not lose its insulating ability when wet—which is one reason why fishermen like it—and can hold up to 30 percent of its weight in water without feeling damp to the touch. Wool fibers have a natural spiral shape which gives them resilience and loft, enabling the fibers to trap thousands of tiny air pockets for insulation. This natural loft serves the same purpose as built-in goosebumps; it is permanent, nearly indestructible and pops back into shape after compression. Wearing a heavy wool sweater is roughly the equivalent of raising the surrounding air temperature 4 degrees.

The fact that wool, like cotton and other natural fibers, absorbs and dissipates body moisture readily— that it "breathes"—is particularly important for winter comfort. The human body loses heat twenty times faster in water than in air, and standing perspiration can draw heat away from the skin like a wick. Silk also breathes well, whereas many synthetic materials act like a big plastic bag and hold dampness against the body or on the inside surface of the material itself, which produces a clammy feeling.

Unfortunately, wool itches. It is also expensive. For those reasons, it is often blended with cotton or a synthetic fiber such as nylon, which not only make wool more comfortable when worn next to the skin, but less expensive and often more durable (if combined with synthetics or silk). Wool long johns can be practically unbearable for some people if they don't have a soft cotton liner woven to the inside. Those allergic to wool won't be helped by such a liner, however.

Cotton is one of the most comfortable clothing materials around. It breathes, and provides good wind protection when tightly woven. In its polished, or smooth, form, it provides relatively little insulation compared to wool, but this can be improved by raising its nap—as in flannel or chamois. Cotton can also be used in pile fabrics such as terrycloth or corduroy, which are warmer than, say, a cotton dress shirt. Unfortunately, cotton feels damp sooner than wool, and loses virtually all of its insulating ability when wet.

As somebody once said, wet wool underwear "isn't like being at home," but it can save your life over cotton on a cold day.

Down, like cotton, loses its insulating value when wet. Otherwise (ignoring costs for a moment), it is wonderful stuff. For lightness and warmth, neither man nor nature has been able to beat it. When magnified, down looks something like a meandering chain of snowflakes, held together by tiny interlocking barbs which form feathery clusters that trap dead air between them. Most down, these days, comes from young ducks and is imported from the People's Republic of China. It takes twelve ducks, or two to three geese, to insulate a down vest. And the cost of that material is rising steadily because supply can't begin to keep up with demand. Down from the eider duck, an endangered species whose natural habitat is north of the Arctic Circle, insulates twice as well as goose down, but is prohibitively expensive. Eiderdown can legally be taken only from abandoned nests, and a comforter may cost several thousand dollars, if it can be found.

Down is a bird's long underwear, worn close to the skin. It is distinguished from feathers by the fact that it doesn't have quills, but in fact most down-filled articles contain at least some feathers and quill fragments. Because of its high cost, it should be used sparingly— in parkas, vests or bedcoverings for example. If the

head and torso are kept warm, hands and feet will follow suit, because their blood supply won't be restricted by the hypothalamus. On the other hand,
mittens worn without a heavy jacket can still result in
cold, numb fingers. Any down-filled coat which extends much below the hips is apt to be more fashionable than practical because cold air can sneak up
underneath it, short-circuiting the down insulation.

Down loses its insulating value when it is wet or
compressed, as when you sit or lie down on it. It
should not be stored compressed or rolled up, and
should be washed as little as possible to preserve natural oils in the plumage. A mild soap is better than de-
tergent here, and the article should be dried at a warm,
not hot, temperature. Drying can take several hours.
Toss an old sneaker in with the garment to help restore
its fluff. If a sleeping bag or parka is merely musty, not
soiled, air it out in sunlight rather than wash it.

Animal fur makes excellent body insulation, but is
too costly to be used economically for that purpose
alone, and is usually worn for its fashion value. Camel
hair and cashmere (taken from Kashmir goats in China
and India) are both much warmer than wool, and
something called Quiviut, which comes from the musk
ox, is more than ten times as warm as wool, per ounce.
Alpaca, llama, and angora fibers are also extraordinar-
ily warm—and expensive. A better bet is to find furs at
yard sales or used-clothing stores and patch them up
or make a new article out of them. Old furs can be re-
stitched with a soft, sturdy facing on the underside for
a superb blanket or comforter. Or it can be used as
trim around hoods and the wrists of a coat, where it
seals out the wind and provides a wonderfully soft,
warm touch on the skin. Artificial, synthetic fur can't
compare with the real stuff, as a rule.

Silk is also a superb insulator, which "breathes" and absorbs moisture without feeling damp. Its high cost and smooth, luxurious feel make it ideal for undergarments worn directly against the skin, including socks, or as a soft liner for wool.

Synthetic materials generally don't insulate as well as natural fibers, and have little or no ability to absorb moisture. Under extremely wet conditions this can be a bonus, as the insulating properties are not affected much by soaking, but in daily living it can hold perspiration against the skin to produce a clammy, claustrophobic feeling. These materials do, however, make excellent windproof and waterproof linings and outer shells, as long as they have some ability to breathe and are separated from the skin by a more absorbent material. The man-made fibers such as acrylic, polyester, nylon and a few others, are relatively inexpensive, easy to care for and durable.

Nylon, in particular, is extremely light and can be woven more tightly than any other fabric—which makes it an excellent windbreaker or windproof shell over down or another good insulating material. It breathes, and is one of the toughest fabrics available for its weight.

Because of its low cost, polyester is edging into the wool and down markets in the form of such insulating materials as Dupont Fiberfill and Damart Thermodactyl. These are easy to maintain and relatively warm (although less so than their natural counterparts), do not absorb water, are inexpensive and dry quickly if soaked.

Now, the trick is to get this stuff arranged on your body in such a way that you stay warm but can still stand up.

The first step is layering. Two thin shirts, socks or

whatever are warmer than a single thick layer of the same material, because they trap a dead air space between them and still air is the basis of all good insulation. Three layers on the torso (say, thermal underwear, a cotton shirt and a wool sweater or jacket) won't restrict movement and can be worn in a house or office in comfort. If this gets too warm, then the thermostat is set too high. Layers also enable you to adjust to quick temperature changes or levels of activity by simply adding or shedding one article.

Five layers is about the maximum for common sense and ease of movement. After that, the insulation value of each additional layer drops drastically by squeezing out the air spaces. They can also restrict blood flow to extremities.

For the same reasons, clothing should also be relatively loose.

Shoes and boots, especially, should provide plenty of room for two pairs of socks without squeezing your toes or being tight on the foot. High heels can restrict blood flow to your heel and toes as quickly as anything, and should be avoided in winter. Clogs with no covering around the heel can also get remarkably cold, although the high wooden sole is good insulation from the cold ground. In general, any foot covering should be able to breathe, because damp socks are the bane of winter comfort. Insulated boots are good for outdoor wear, and the variety with rubber lowers and leather uppers are especially comfortable in being waterproof where it counts, but able to breathe away moisture. Insulating innersoles, available at drugstores, department stores and shoe repair shops, can triple the comfort of almost any boot. Whenever possible, tuck your trousers or slacks into the boot tops; it keeps cold air from

getting to places you don't want it, and keeps legs considerably warmer.

For socks, virgin (not used, or reprocessed) wool
with the lanolin still in it is ideal. A thinner, softer sock
can be worn underneath. If you can find, or make,
socks with a double layer below the foot and a single
layer above, these will be unusually warm and comfortable. Down-filled slippers are cozy for use around
the house, and don't cost that much. Leg warmers, the
type worn by dancers to avoid cramps, are excellent to
wear beneath skirts or dresses and have become fashionable—and therefore easier to find.

Thermal underwear, meanwhile, is the best investment of all. It provides warmth and coziness with little
or no bulk, and can be worn under almost anything. A
double layer of wool and cotton is ideal, but many
other fabrics do nearly as well—as long as they can
breathe. Union suits these days can be found in almost
anything from synthetic fibers to angora, cashmere and
even down (which is somewhat bulky). Waffle knit or
net underwear is popular and practical because the
small air spaces between the weave provide insulation.
However, these work best when worn under another
layer of long johns, to keep the dead air spaces intact.
As long as the material has good insulating capacity,
buy whatever feels the most comfortable. It's wise to
save heavy pairs for outdoors, and lighter, softer material for use inside and at the office.

The best protection for hands is a pair of mittens.
Unlined leather gloves provide very little protection
against the cold, and in no case should gloves fit so
tightly that they restrict circulation. Mittens of down
(on the backs of the hands) or with wool inserts are
about as warm as anything, because they not only in-

sulate but keep each finger warming the next. It's the old principle of reducing surface area. Fingerless gloves are also popular again, because they keep your hands warm while allowing free finger movement. They look a bit odd at first, but before long you can forget you have them on. Or so says my wife, who wears them almost everywhere but in the bath.

For practical head and neck warmth, the ideal solution is a heavy sweater with an attached hood. This becomes a hat, scarf and sweater in one; costs less than buying the three garments individually and can be worn inside or outdoors (as a coat). Turtleneck sweaters also provide good neck protection, as do the collars of wool shirts. Outside, a scarf is best. Part of the trick to dressing warmly indoors is to do so without looking or feeling like a snowman, which is why multipurpose garments are useful. Hats are being worn indoors again, so get used to wearing one. Wool stocking caps, or "watch" caps as they are sometimes called, are extremely practical, but can be itchy on the forehead.

One of the most important parts of the body to keep warm is the lower back where the kidneys and liver are located, so keep this area protected. Vests and sleeveless sweaters work well. When buying wool or flannel shirts, make sure they fit relatively snugly here, and have tails long enough to tuck in so they don't pull out everytime you bend over.

When buying a parka, get one with down or other insulating material in the collar and hood, and a snug fit around the wrists, hips and neck. Also look for a good fill under the armpits, and a front opening that doesn't allow wind to penetrate through the zipper. An extra flap is good here. Plastic zippers tend to freeze less quickly than metal ones. Any material in a coat or parka that flaps around in the breeze below your waist does little good from •an insulating standpoint. Coats can be both fashionable and warm, so look for both. If the coat or parka has a hood, make sure this covers your head without blocking vision, and can be drawn snugly around the face. Open-front floppy hoods only help slow down a tail wind—and have a tendency to fill up with snow or cold rain just before you put them over your head.

The most frustrating part of all this can be the fact that some articles of winter clothing, such as fingerless gloves, are virtually impossible to find in women's sizes; or the women's version is less well insulated than the same article made for men. So it's often smart to look for a small man's size, rather than go through winter stylish but shivering. Maybe this will change.

Meanwhile, a slew of other cold weather clothing ideas are coming on the market—or coming *back* on the market—so look around for things you may not have thought of before. Even the wool chemise (feminine for "undershirt") is being sold in scattered stores again, and is worth picking up if you can find one.

Who knows? Bonwit Teller might stock them next year.

16

The Bed

≈ The practice of staying snug and warm on a cold
≈ winter night by curling up with pets on the bed is
an old custom in parts of Europe. A two-dog night, of
course, is colder than a one-dog night, but not as cold
as a three-dog night—which apparently is a serious
thing to live through, because a four-dog night has
never been recorded. The heat from three dogs is all
you will ever need to survive until dawn, and that
works out nicely, because with more dogs than that
you would have to sleep on the floor, which is worse
than where you started from.

Now, spending the night with just *one* of our dogs
would be like curling up in the puckerbush next
to a large bear with digestive problems. Actually, it
wouldn't be quite that nice, but close enough. It is not
a pretty thought. The people who spend multidog
nights must have dogs who never wallow in barnyards
or burr thickets, or trot in the kitchen door carrying
strange, limp objects.

At any rate, a two-dog night is considered optimal
for comfort because an animal can curl up against
either side of your body. (An odd number works
badly—the strong side will eventually push you off the
bed, and you're right back there on the floor again.)
Two dogs also keep you healthy, assuming they sleep

on top of the blankets, as this pins you tightly to the bed and precludes any possibility of catching a chill while trying to reach the bathroom at night. You must never turn over or curl up, or this can dislodge the dogs, resulting in low growls on either flank. Moreover, the problem of trying to sleep late while being closely scrutinized by large animals has never been worked out. Few noises are louder than the silent stare of an anxious dog.

This is rather pleasant business, though, when weighed against the problems of sleeping with other animals, such as unhousebroken goats. You haven't *lived* until you have spent a one-goat night.

Guenevere, one of our young Toggenburgs, was tossed about pretty badly by a dog one late fall day, which put her neck into something of an S-turn. Nothing was broken, but the poor girl was miserable, and said so. My wife, who feels an enduring softness and understanding for all creatures alive (or recently so), declared she would sleep downstairs with Guenevere that night to help comfort the animal. But I objected. It is a hard thing to watch a mature adult pout, plead and grow close to tears without giving in to it—but my wife said such carryings-on would do me no good at all; she was staying with the goat, and that was that.

She spread an old horse blanket on the floor beside the couch, dug out her sleeping bag, mixed a Bloody Mary and then settled in for the night. Unfortunately, the goat edged up on to the couch at some point and began working her way toward the celery stick in the still-untouched Bloody Mary. My wife awoke to find Guenevere standing over her head, drinking. She let out a scream. Guenevere let out a tide of nanny-berries. By that time, the Bloody Mary was gone, and the goat spent the rest of the night falling off furniture while searching for more vodka. It wasn't funny.

As a rule, blankets work better on the winter bed than dogs, goats, wounded hawks, squirrels or any of the other wildlife which at various times has found its way into our bedroom. Fur comforters are fine; they just shouldn't be moving.

Warm coverings are especially important in bed, not only because you may spend a third of your life there, but because the body's metabolism—and temperature—drops during this extended period of inactivity. (One study shows this may not be true of Eskimos.)

Many of the same rules which apply to dressing warmly by day also apply at night. That is, use several layers of materials that can breathe and will provide insulation.

Sheets serve a dual purpose of feeling smooth next to the skin and keeping blankets clean. Cotton is the traditional material for sheets, and is still hard to beat. Linen and even silk were once popular, but price has pretty much driven these off the market. Muslin (a coarse cotton weave) and percale (a fine, closely woven cotton) have also been displaced to a great extent by polyester, or polyester-cotton blends. These are durable, and require little ironing, but don't breathe as well as pure cotton. Or feel the same.

Cotton flannel sheets are widely available now, and it is worth buying a few. They are reasonably priced, have insulating value, and are as comfortable next to your skin on a cold winter night as anything you will find. Almost anything. A single flannel top sheet is usually enough to eliminate the shock of entering a cold bed. They are sturdy and machine-washable, and don't require ironing.

Above that, a heavy wool blanket is still the best for warmth and security—and the so-called Hudson's Bay blanket is the warmest thing I have ever slept under. These are expensive, but they can be handed down to your grandchildren. Two good wool blankets is all you should ever need. They are heavy, for one thing, and a third blanket adds virtually no insulation because it squeezes out the air layer which provides much of the insulation in bed coverings.

Down comforters, or even an unzipped sleeping bag, make especially warm blankets, but they tend to slide off the bed and end up in a pile on the floor—which the dog quickly claims for its own. So-called "space blankets," often containing a thin aluminum film sheathed in Mylar, reflect heat back to your body, but hold moisture in at the same time.

An electric blanket is another wise investment for the winter bed. It isn't apt to need more than $8 worth of electricity per heating season, and serves the same function as an early American bed warmer, taking the chill off the sheets and blankets before you have to jump in. The human body uses a good deal of heat (and misery) to warm a cold bed, but requires very little to keep it that way. In fact, you will stay much warmer under a single wool blanket than if you were clothed in the same material, because in bed very little body surface area is exposed to the cold—especially if you curl up. Each part of the body is in the same snug cocoon, undivided by belts, collars or wrist cuffs. Those body sacks which have become so popular work on the same principle. Moreover, blood flow to your extremities is better when lying down.

To get the most mileage out of an electric blanket, set it on "low" a half-hour or so before retiring, rather than cranking it up to "hot" or "sear" after you have hit the sheets. This will save electricity in the long run, and will save you from going through the delicate contortions of trying to get into a cold bed without actually touching the icy sheets. (There is another way, but it requires a sleeping partner and is difficult to describe without slides. For beginners, the best bet is to settle all arguments beforehand, then, just before climbing under the covers (you might actually want to *lift* the covers to show good intentions), remember you left the back door unlocked. Or left the iron on. Whatever—your partner will, in most cases, begin warming the sheets without you.)

If such a bed warmer is not available, or the electric blanket has been shredded by fighting dogs, a heating pad or hot water bottle will do. Or a thin slab of soapstone can be warmed by the fire, wrapped in newspaper and a towel and slipped between the sheets. It will give off heat for at least an hour, but make sure it isn't too hot to hold. Early American bed warmers consisted of a round metal pan with a lid, attached to a long wooden handle. Fireplace coals and warm ash were put inside, and then the bed warmer was passed under the covers for a few moments. Many of these, made of brass or tin, are not only practical but decorative, and can be hung by the fire when not in use. Be careful, though—the metal bottoms are often worn thin or rusted.

The bed itself can also help keep you warm. In years past, beds tended to be higher than they are now—in part so that a child's trundle bed could be rolled out of the way underneath. But height also keeps sleepers well above cold floor drafts. A skirt or valance around the bed's feet served the same purpose, and four-poster canopy beds are a vestige of the days when the night winter bed was totally enclosed behind drawn curtains, for a snug, warm and private sanctuary. The posts remain today, and sometimes the canopies, but the rest is gone.

Koreans prefer sleeping on the floor because of an ingenious system they have in traditional houses of using the floor itself as part of the stove chimney. The stove or oven is placed below the level of the floor, which is built of hard, adobe-like earth with open channels under it. When the stove is fired up, heat drifts through these channels, warming the floor before it is vented from the house.

For the body itself, a cotton flannel nightgown or pair of pajamas is unbeatable for price, warmth and comfort. In either case, pajamas or gown, it should have a high neck, long sleeves and a bottom hem that just about touches the floor when you are standing. Nightshirts are fashionable but drafty. Many people are reverting to nightcaps and night socks (made especially for the purpose), or even fingerless gloves or mittens, for sleeping or reading in bed on the coldest winter nights.

Unfortunately, children's sleepwear is more limited. In this country, nearly all of it is made of polyester these days because this synthetic fiber is more flame-proof than cotton. Since 1972, federal regulations have required that children's sleepwear either burn slowly or be self-extinguishing, and cotton meets these standards only if treated with chemical flame retardants—the most popular of which (Tris) has been the subject of health controversy and banned from use because of its possible carcinogenic properties. Polyester, however, provides little insulating value compared to cotton flannel, and does not breathe away moisture from the skin.

The best bet is to use hand-me-down cotton pajamas

made prior to 1972, and patch them up if necessary. Flannel pajamas, like old wool shirts, don't even begin to get comfortable until they have been patched a few times. However, avoid filmy, frilly, flowing gowns, which are not only hazardous if a child wanders too near a fire, but provide little protection from the cold anyway. If I were chilly, I'd get near a fire too. Let them wear pajamas in bed, not around the house. There aren't as many fire hazards in a crib.

In any case, it is wise to keep a screen or other physical barrier between children and a space heater or any open-flame fire. Low gates, of the type used on stairways, are ideal for this purpose because they keep children away from wood stoves or fireplaces without blocking heat. If the gate is wood, keep it the same distance from a stove as any other combustible material. A simple and handsome barrier can be made of 4-by-4-inch posts and 1-by-4-inch boards, with a gate in front. The idea is to provide not only a physical barrier, but a psychological one, beyond which children can be taught not to go. Children love to imitate adults, and adding things to a fire is particularly tempting for them.

The lack of warmth in polyester sleepwear can be compensated for with a good quilt or blanket, but if you prefer not to use this material, and cotton hand-me-downs are not available, it is easy to make a cozy sleeping "sacque" from scratch. This is a sort of bundle bag with sleeves, and has a drawstring at the bottom which can be closed against the cold, but allows easy access for changing young children's diapers. It is practical in winter, easier to sew than pajamas and takes a good while to grow out of.

Of course, the simplest and most rewarding way of all to keep young children warm and safe on a winter's night is to let them sleep with each other. Many parents even share a family bed, or keep the crib in their room, because they say it alleviates bedtime crying, infants sleep better and have fewer (if any) nightmares and everyone stays warmer. Nursing mothers often like the arrangement, not only for convenience but because it fulfills a sometimes overpowering desire to be with their babies. Toddlers are allowed to climb in the bed when, in one way or another, they get cold. Children are gone soon enough.

While this is commonplace in many countries, and "bundling" is woven into our own historic fabric, such practices are discouraged in our society now. Today we make children sleep alone from the moment of birth, and, to simulate the warmth and comfort of a parent or sibling, fill the crib in the next room with all manner of odd devices—pacifiers, stuffed animals, ticking clocks, tape recordings of a human heart and now even little windup trains that run along the crib railing. ("It goes back and forth so you don't have to," the manufacturer says.) We instruct new mothers not to rush in and pick up their crying infants, when the instinct to do so wells up from the very depths of motherhood—and the notion of a separate bed and room for each child has evolved into a sort of national birthright and sign of status, which we seem determined to enforce at any price. All the while we wonder where the close family went.

Now, there may be times when you wish each child had a separate *dwelling*, never mind a separate room, but children themselves don't necessarily agree. One couple whose youngsters were accustomed to sharing beds recall moving to a new town and having their children come home from playing at a neighbor's house one afternoon. One of them asked if the kids next door each had a separate room because they were being punished.

In any case, a shared bed keeps you warm in many ways, and the practice is apparently gaining favor again as room temperatures drop. There are problems with all of this, some trivial, others not, and others which turn out not to be problems at all (yes, children could get in the habit of seeking out someone they love), but the real difficulty is in having the self-assurance to decide what makes good sense and is right in this house.

We live in an age of experts now, without whose high-grade counsel we cannot be trusted to dress ourselves alone, make love, run, grow vegetables, give birth, raise children or even . . . well . . . keep warm in winter.

That's 12 inches of insulation in the attic, dammit, not 10 or 15. Get it right.

We have let our instincts atrophy, no longer trusting them, and seen wisdom born of experience grow cold between the lines of advice columns. We wait for the weather forecaster to tell us what we already know—that the bitter cold will break tomorrow—because we've seen a haze around the moon and know which way the wind is blowing, if only we'd listen. We know, without reading about it, that oak burns longer than pine, and that heat rises; that the January sun will heat a car through its closed windows, and do the same for a house. We have felt cold air roll off a windowsill, and registered that knowledge; and we have known, without asking, that winter is close at hand now, because mosquitoes have left the marshes and Canada geese are flying. With our own hands we have built houses that stand for three hundred years. We have cut wood, grown vegetables, taken responsibility for our own families, and—by sifting technical expertise through the strainer of our own acquired wisdom—we have done quite well, thank you.

Anyway, as to the matter of young children, it doesn't hurt to sleep with them. They rarely have fleas, and don't growl at night.

17

Glögg

≈ In Scandinavian countries, there remains an old
winter holiday custom of brewing up a special
demon nectar called glögg (pronounced "gleug" or
"glug," depending on who you ask and how many
they've had), which is mulled wine. Traditionally, each
family has its own recipe, and a kettle of it is kept
brewing at the back of the stove throughout the
Christmas season. Cully the Swede who runs the fish
market in town and serves ten gallons of glögg to his
customers each Christmas Eve, inherited one of the
finest recipes around, but doesn't get too specific about
it. Says it has something to do with 190-proof alcohol
and federal law. It takes the chill off a winter night as
nothing else can, and the mere fragrance of it brewing
in an iron kettle can cheer up an entire house.

The basic ingredients are dry red wine and a strong
vodka or aquavit. To this, add cardamom seeds
(slightly crushed), cinnamon sticks, orange peel, rai-
sins, dried apples or other fruit and maybe a few
cloves. Ingredients are not precise because there is no
single recipe for this drink. It is a hallmark of the winter kitchen, and should be mixed to suit your taste.
However, cups of wine should outnumber vodka by
maybe ten to one.

And use a dry wine. It is easy to make this drink too

sweet and trim its character that way. Some people add angostura bitters, vermouth and sometimes a cup of sugar to a six-quart batch. A cup or two of blanched almonds doesn't hurt, and a piece of fresh ginger adds a good taste.

The mixture should simmer and then stand for at least twelve hours before serving; some people like to let it gather strength for a month or so. The flavors mingle and wonderful things happen. Glögg should never be boiled because the alcohol will be the first thing to evaporate. In fact, that principle is what makes a whiskey still work.

Or that's what Cully the Swede tells me.

Suggested Further Reading

Cole, John N., and Charles Wing. *From the Ground Up*. Boston: Atlantic–Little, Brown, 1976.

Gay, Larry. *Central Heating with Wood and Coal*. Brattleboro, VT: Stephen Greene Press, 1981.

McCullagh, James C., ed. *A Solar Greenhouse Book*. Emmaus, PA: Rodale Press, 1978.

Olkowski, Helga, Bill Olkowski, Tom Javits and the Farallowes Institute. *The Integral Urban House: Self-Reliant Living in the City*. San Francisco: Sierra Club, 1979.

Rothchild, John. *Stop Burning Your Money*. New York: Random House, 1981.

Sheltons, Jay, and Andrew Shapiro. *The Woodburner's Encyclopedia*. Waitsfield, VT: Crossroads Press, 1976.

Wing, Charles. *From the Walls In*. Boston: Atlantic–Little, Brown, 1978.